NIST Special Publication 800-76-2

Biometric Specifications
for Personal Identity Verification

Patrick Grother
Wayne Salamon
Information Access Division
Information Technology Laboratory

Ramaswamy Chandramouli
Computer Security Division
Information Technology Laboratory

July 2013

U.S. Department of Commerce
Penny Pritzker, Secretary

National Institute of Standards and Technology
Patrick D. Gallagher, Under Secretary of Commerce for Standards and Technology and Director

Certain commercial entities, equipment, or materials may be identified in this document in order to describe an experimental procedure or concept adequately. Such identification is not intended to imply recommendation or endorsement by NIST, nor is it intended to imply that the entities, materials, or equipment are necessarily the best available for the purpose.

There may be references in this publication to other publications currently under development by NIST in accordance with its assigned statutory responsibilities. The information in this publication, including concepts and methodologies, may be used by Federal agencies even before the completion of such companion publications. Thus, until each publication is completed, current requirements, guidelines, and procedures, where they exist, remain operative. For planning and transition purposes, Federal agencies may wish to closely follow the development of these new publications by NIST.

Organizations are encouraged to review all draft publications during public comment periods and provide feedback to NIST. All NIST Computer Security Division publications, other than the ones noted above, are available at http://csrc.nist.gov/publications.

Comments on this publication may be submitted to:

National Institute of Standards and Technology

Attn: Information Access Division, Information Technology Laboratory

100 Bureau Drive (Mail Stop 8940) Gaithersburg, MD 20899-8940

Reports on Computer Systems Technology

The Information Technology Laboratory (ITL) at the National Institute of Standards and Technology (NIST) promotes the U.S. economy and public welfare by providing technical leadership for the Nation's measurement and standards infrastructure. ITL develops tests, test methods, reference data, proof of concept implementations, and technical analyses to advance the development and productive use of information technology. ITL's responsibilities include the development of management, administrative, technical, and physical standards and guidelines for the cost-effective security and privacy of other than national security-related information in Federal information systems. The Special Publication 800-series reports on ITL's research, guidelines, and outreach efforts in information system security, and its collaborative activities with industry, government, and academic organizations.

Abstract

Homeland Security Presidential Directive HSPD-12, Policy for a Common Identification Standard for Federal Employees and Contractors [HSPD-12], called for Homeland Security Presidential Directive HSPD-12, Policy for a Common Identification Standard for Federal Employees and Contractors [HSPD-12], called for new standards to be adopted governing interoperable use of identity credentials to allow physical and logical access to Federal government locations and systems. The Personal Identity Verification (PIV) standard for Federal Employees and Contractors, Federal Information Processing Standard Personal Identity Verification (PIV) of Federal Employees and Contractors (FIPS 201), was developed to define procedures and specifications for issuance and use of an interoperable identity credential. This document, Special Publication 800-76 (SP 800-76), is a companion document to FIPS 201. It describes technical acquisition and formatting specifications for the PIV system, including the PIV Card itself. It also establishes minimum accuracy specifications for deployed biometric authentication processes. The approach is to enumerate procedures and formats for collection and preparation of fingerprint, iris and facial data, and to restrict values and practices included generically in published biometric standards. The primary design objective behind these particular specifications is to enable high performance and universal interoperability. The introduction of iris and face specifications into the current edition adds alternative modalities for biometric authentication and extends coverage to persons for whom fingerprinting is problematic. The addition of on-card comparison offers an alternative to PIN-mediated card activation as well as an additional authentication method.

Keywords

biometrics; credentials; identity management

Acknowledgements

The authors from the National Institute of Standards and Technology (NIST) wish to thank their colleagues who reviewed drafts of this document and contributed to its development. Particular thanks go to the many external commenters who produced detailed comments on the drafts, to Charles Wilson who directed the development of the original SP 800-76 and its early update, SP 800-76-1, and to R. Michael McCabe for his extensive knowledge of various fingerprint standards and the Federal Bureau of Investigation's procedures. The authors also gratefully acknowledge and appreciate the many contributions from the public and private sectors for the continued interest and involvement in the development of this publication.

Executive Summary

Scope: Homeland Security Presidential Directive 12, *Policy for a Common Identification Standard for Federal Employees and Contractors* [HSPD-12], called for new standards to be adopted governing interoperable use of identity credentials to allow physical and logical access to Federal government locations and systems. The Personal Identity Verification (PIV) standard for Federal Employees and Contractors, Federal Information Processing Standard 201, *Personal Identity Verification (PIV) of Federal Employees and Contractors* (FIPS 201), was developed to define procedures and specifications for issuance and use of an interoperable identity credential. This document, Special Publication 800-76 (SP 800-76), is a companion document to FIPS 201. It describes technical acquisition and formatting specifications for the PIV system, including the PIV Card[1] itself. It also establishes minimum accuracy specifications for deployed biometric authentication processes.

Approach: The approach is to enumerate procedures and formats for collection and preparation of fingerprint, iris and facial data, and to restrict values and practices included generically in published biometric standards. The primary design objective behind these particular specifications is to enable high performance and universal interoperability. The introduction of iris and face specifications into the current edition adds additional modalities for biometric authentication and extends coverage to persons for whom fingerprinting is problematic. The addition of on-card biometric comparison offers an alternative to PIN-mediated card activation as well as an additional authentication method. For the preparation of biometric data suitable for the Federal Bureau of Investigation (FBI) background check, SP 800-76 references the ANSI/NIST Fingerprint Standard [AN2011] and the FBI's Electronic Biometric Transmission Specification [EBTS].

Domain of use: Recognizing that PIV specifications are sometimes leveraged in other identity management applications, it should be noted that derivative programs should adopt these PIV-specifications with appropriate deliberative technical augmentation. The biometric data elements contained in this standard are suitable for one-to-one verification of document holders when other application-specific factors are maintained. But, for example, if the fingerprint templates mandated here are used in conjunction with fingerprints captured on non-PIV compliant fingerprint sensors, there may be systematic degradations in recognition accuracy. Similarly, while it would be appropriate to take the compressed iris specification defined here for use on a nation state's e-Passports[2], it would be technically suboptimal to then copy those iris images to be the enrollment samples of an expedited traveler program running in one-to-many single-factor mode (i.e., the mode of [NEXUS,UKIRIS]).

Biometric data used outside the PIV Data Model is not within the scope of this standard.

---

[1] A physical artifact (e.g., identity card, "smart" card) issued to an individual that contains a PIV Card Application which stores identity credentials (e.g., photograph, cryptographic keys, digitized fingerprint representations) so that the claimed identity of the cardholder can be verified against the stored credentials by another person (human readable and verifiable) or an automated process (computer readable and verifiable).

[2] In the Data Group 4 container defined for iris data by [ICAO].

Table of Contents

List of Figures

List of Tables

1. Introduction

1.1 Authority

This document has been developed by the National Institute of Standards and Technology (NIST) in furtherance of its statutory responsibilities under the Federal Information Security Management Act (FISMA) of 2002, Public Law 107-347.

NIST is responsible for developing standards and guidelines, including minimum requirements, for providing adequate information security for all agency operations and assets, but such standards and guidelines **shall** not apply to national security systems. This recommendation is consistent with the requirements of the Office of Management and Budget (OMB) Circular A-130, Section 8b(3), Securing Agency Information Systems, as analyzed in A-130, Appendix IV: Analysis of Key Sections. Supplemental information is provided in A-130, Appendix III.

This recommendation, prepared for use by federal agencies, may be used by non-governmental organizations on a voluntary basis and is not subject to copyright. Nothing in this document should be taken to contradict standards and guidelines made mandatory and binding on Federal agencies by the Secretary of Commerce under statutory authority. Nor should this recommendation be interpreted as altering or superseding the existing authorities of the Secretary of Commerce, Director of the Office of Management and Budget, or any other Federal official.

1.2 Purpose and scope

FIPS 201 [FIPS], Personal Identity Verification (PIV) for Federal Employees and Contractors, defines procedures for the PIV lifecycle activities including identity proofing, registration, PIV Card issuance and re-issuance, chain-of-trust operations, and PIV Card usage. [FIPS] also defines an identity credential that includes biometric data. Requirements on interfaces are described in [800-73, parts 1-3]. Those on cryptographic protection of the biometric data are described in [FIPS] and in [800-78].

This document contains technical specifications for biometric data mandated or allowed in [FIPS]. These specifications reflect the design goals of interoperability, performance and security of the PIV Card and PIV processes. This specification addresses iris, face and fingerprint image acquisition to variously support background checks, fingerprint template creation, retention, and authentication. These goals are addressed by normatively citing and mandating conformance to biometric standards and by enumerating requirements where the standards include options and branches. In such cases, a biometric profile can be used to declare what content is required and what is optional. This document goes further by constraining implementers' interpretation of the standards. Such restrictions are designed to ease implementation, assure conformity, facilitate interoperability, and ensure performance, in a manner tailored for PIV applications.

The biometric data specifications herein are mandatory for biometric data carried in the PIV Data Model (Appendix A of [800-73, Part 1]). Biometric data used outside the PIV Data Model is not within the scope of this standard.

This document does however specify that most biometric data in the PIV Data Model **shall** be embedded in the Common Biometric Exchange Formats Framework [CBEFF] structure of Section 9. This supports record integrity (using digital signatures) and multimodal encapsulation.

This document provides an overview of the strategy that can be used for testing conformance to the standard. It is not meant to be a comprehensive set of test requirements that can be used for certification or demonstration of compliance to the specifications in this document. NIST Special Publications 800-85A and 800-85-B [800-85] implements those objectives.

1.3 Audience and assumptions

This document is targeted at Federal agencies and implementers of PIV systems. In addition, it should be of interest to the biometric access control industry. Readers are assumed to have a working knowledge of biometric standards and applications.

1.4 Overview

1.4.1 Document structure

This document defines:

— In Section 2, acronyms and terms;

— in Section 3, the fingerprint acquisition process, requirements for transmission of data to FBI, and a format for agency-optional image retention;

— in Section 4, the format of the PIV Card minutiae templates for off-card authentication, and specifications for algorithms used in the generation and matching of such;

— in Section 5, the formats and data structures for minutiae used in on-card comparison operations, and specifications for algorithms used in the generation and matching of such;

— in Section 6, the format for iris data stored on and off PIV Cards, and specifications for cameras and algorithms used for the collection, preparations and matching of such;

— in Section 7, the format and data structures for facial images on PIV Cards, and specifications for collection thereof;

— in Section 8, interface specifications for biometric sensors;

— in Section 9, the CBEFF header and footer supporting digital signatures on all PIV biometric data;

— in Section 10, minimum accuracy specifications;

— in Section 11, additional conformance information, beyond the specifications embedded in Sections 4 through 7;

— in Section 12, references.

Figure 1 gives an approximate procedure for biometric data acquisition and disposition.

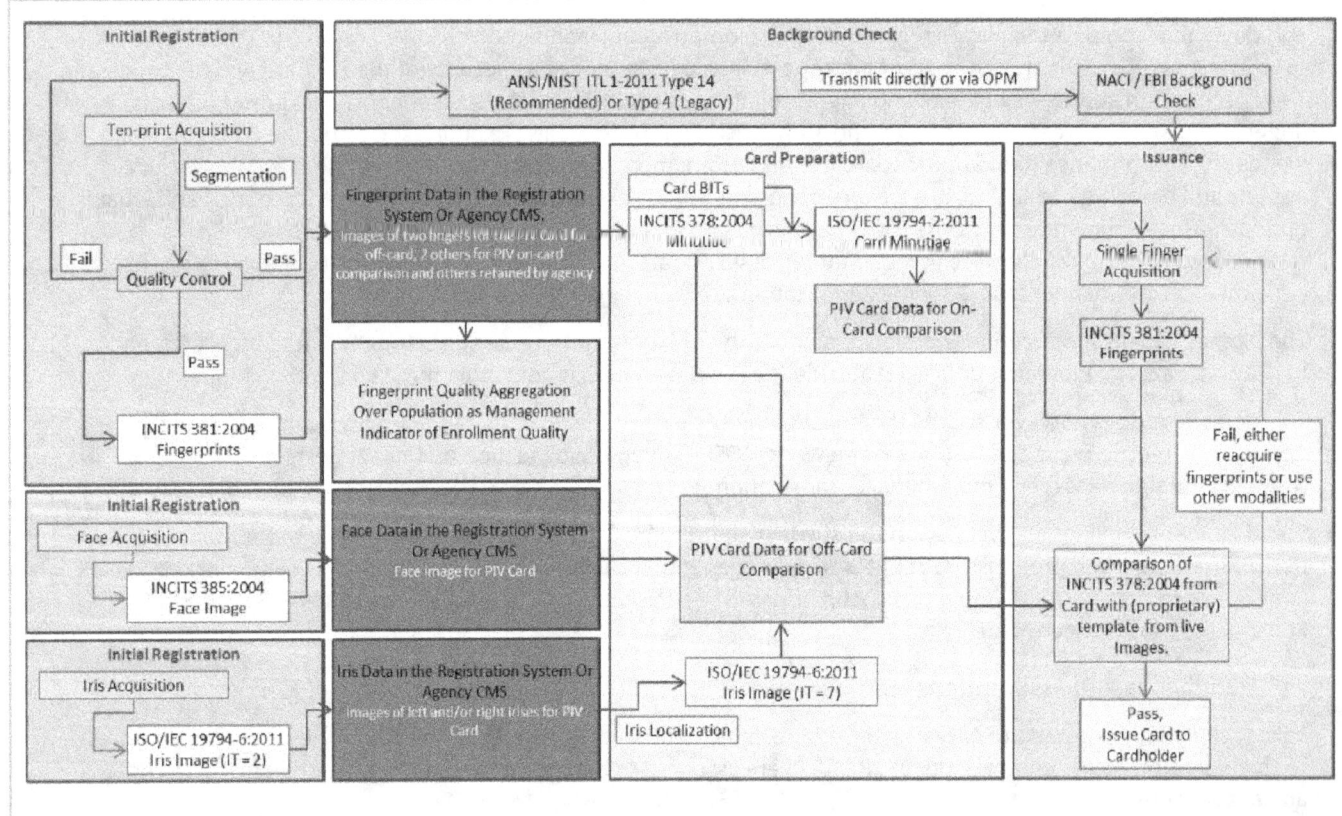

Figure 1 – PIV biometric data flow

1.4.2   Inclusion of iris recognition

Iris specifications are included, in Section 6, to support biometric authentication of individuals. [FIPS] allows use of iris for this purpose. The recommendation to agencies to install and operate iris equipment in its PIV issuance processes allows agencies to additionally populate PIV Cards with iris as an alternative authentication modality. [FIPS] requires the cardholder to enter a PIN number to release the templates. When the card is cryptographically authenticated, this constitutes multi-factor authentication.

1.4.3   Inclusion of fingerprint on-card comparison

[FIPS] requires fingerprint templates of Section 4 as the mandatory biometric element for PIV. These templates are intended to be compared on a reader device with templates collected in an authentication attempt. [FIPS] requires the cardholder to enter a PIN number to release the templates. When the card is cryptographically authenticated, this constitutes multi-factor authentication.

Agencies may additionally choose to populate the card with an on-card comparison algorithm, and on-card comparison templates. The specifications for these appear in Section 5. [FIPS] does not require PIN entry ahead of a fingerprint minutiae on-card comparison transaction. Indeed, [FIPS] extends on-card comparison *as an alternative* to PIN entry in altering the security state of the PIV Card.

Agencies should consider economics of on vs. off card comparison. Particularly, a security flaw, for example, in a card may have different remediation cost than in a card reader.

Table 1 describes the differences between the off-card and on-card specifications.

Table 1 – Summary of properties and roles of on- and off-card fingerprint comparison

#	Aspect	Off-card comparison	On-card comparison
1.	[FIPS] requirement on presence of biometric data	Mandatory	Optional
2.	Use cases and pre-requisites for access to the data	See [FIPS]	
3.	Interagency interoperable	Yes	No (As OCC is optional, it is only interoperable across agencies if both agencies implement it).
4.	Number of fingers required to be stored on card	2. But 0 or 1 are allowed in exceptional cases – see [FIPS]	1 or 2
5.	Number of fingers to be used in a biometric operation	1 or 2	1 or 2
6.	Which fingers	Members of the set A, which is a subset of the ten finger set T	Members of the set B, which is a subset of the ten finger set T, and $|A \cap B| \geq 0$, i.e., , fingers may be the same. See Section 5.4.
7.	Encoding of specific fingers	INCITS 378:2004 [MINUSTD]	ISO/IEC 7816-11:2004 [CARD-BIO]
8.	Data format specifications	This document, Section 4	This document, Section 5
9.	Card interface specifications	SP 800-73-4 [800-73]	SP 800-73-4
10.	Underlying data format standard	INCITS 378:2004 [MINUSTD]	ISO/IEC 19794-2:2011 [CARD-MIN] This template **shall** be computed from the off-card INCITS 378:2004 template.
11.	Fingerprint capture device for biometric operations	Plain impression as specified in Section 4.7	
12.	Accuracy testing	MINEX III	MINEX IV

1.5   Relation to other biometric applications

[FIPS] advances a PIV concept of biometric operations that is three-factor[3]: A PIN verification is required before biometric data is read from the PIV Card and matched in a 1:1 mode during authentication. In other programs, biometrics are sometimes stored on a central server, or read from a card and cached on one. In others, the biometric is matched in a one-to-many mode without presentation of a card to claim an identity. There are tradeoffs with such approaches.

---

[3] NIST Special Publication 800-63-1 Electronic Authentication Guideline identifies three factors - things a person has, knows, and is.

- PIV Card read times are replaced with network transmission times.
- PIN entry times are eliminated but the something-you-know additional factor is lost.
- The remote server is subject to physical or logical attack (e.g., a denial-of-service attack). Many kinds of templates stored on a server can be reversed to produce a matchable-sample [REVFING, REVIRIS, REVFACE]. Formal template protection schemes, which mitigate the effect of compromise of a database, require further testing.
- If the PIV Card is not used to make an explicit claim to one of the N enrolled identities, the biometric-only one-to-many authentication loses the something-you-have factor, and necessitates mitigation of a N-fold increase in false match rates.
- Such use cases are not addressed by this specification.

1.6   Second generation standards

Since the first publication of SP 800-76 in 2005, considerable effort has been dedicated to the development of second-generation biometric data interchange standards. These are the various parts of ISO/IEC 19794, as revised, and updates to the analogous INCITS standards in c. 2009. These standards have not been adopted here as replacements for the extant PIV biometric standards - INCITS 385:2004 (face), INCITS 381:2004 (fingerprint image), and INCITS 378:2004 (fingerprint minutiae) – because

- they are not binary compatible with the earlier standards,
- deployed infrastructure (readers) would need to be updated to support both the legacy and second generation standards, and
- they confer essentially no performance advantages over the earlier standards.

For data elements that were not included in the 2005 PIV specifications, i.e., images to support iris recognition and minutia data to support on-card comparison, this specification adopts the latest ISO/IEC 19794 Part 2 and Part 6 standards [CARD-MIN, IRISSTD].

2.1 Terms

Term	Definition
Fingerprint segmentation	Segmentation is the automated (and often manually reviewed) separation of an image of N fingers into N images of individual fingers. N is usually four, for the index through little finger, and two for a capture of two thumbs.
Iris segmentation	Segmentation is the automated (and possibly manually reviewed) detection of the iris-sclera and iris-pupil boundaries. This localizes the iris texture that is used for actual recognition.
One-to-many	Of or relating to biometric identification in which submitted feature data is compared with that of all enrolled identities.
One-to-one	Of or relating to biometric verification in which submitted feature data is compared with that of one, claimed, identity.

Other biometrics-related terms are defined and harmonized in [VOCABSTD].

2.2 Acronyms

Acronym	Definition
BIT	Biometric Information Template – a [CARD-BIO] data structure indicating Card capability
CBEFF	Common Biometric Exchange Formats Framework
DET	Detection Error Tradeoff (characteristic) – A plot of FRR vs. FAR, or FNMR vs. FMR, used to inform security-convenience tradeoffs in (biometric) authentication processes
FAR	False Accept Rate (defined over an authentication transaction)
FIPS	Federal Information Processing Standard
FMR	False Match Rate (defined over single comparisons)
FNMR	False Non-Match Rate (defined over single comparisons)
FRR	False Reject Rate (defined over an authentication transaction)
FTE	Failure to Enroll Rate
EBTS	Electronic Biometric Transmission Specification (See References Section 12)
JPEG	Joint Photographic Experts Group, standardized compression algorithm for face images
OCC	On-card comparison
PNG	Portable Network Graphics, standardized lossless compression algorithm for images
IREX	Iris Exchange – the NIST program supporting iris-based biometrics
MINEX	Minutia Exchange – the NIST program supporting minutia-based biometrics
NACI	National Agency Check with Inquiries
NFIQ	NIST Fingerprint Image Quality – an automated algorithm for quantifying good fingerprint images; available as open-source.
SP	Special Publication – a designation for NIST documents, sometimes supporting FIPS.
WSQ	Wavelet Scalar Quantization

2.3 Organizations

Acronym	Definition
ANSI	American National Standards Institute
FBI	Federal Bureau of Investigation
IEC	International Electrotechnical Commission
INCITS	InterNational Committee for Information Technology Standards
ISO	International Organization for Standardization
ITL	Information Technology Laboratory (of NIST)
NIST	National Institute of Standards and Technology
OMB	Office of Management and Budget
OPM	Office of Personnel Management
SC 37	The Biometrics standardization subcommittee under ISO/IEC Joint Technical Committee 1

3.1   Scope

The specifications in this Section pertain to the production of the mandatory PIV biometric enrollment data. That is, this Section provides specifications for acquisition, formatting, and storage of fingerprint images and templates. The following is an overview of the material covered in this Section.

- Sub-Section 3.2.1 gives specifications capture of fingerprints for PIV Registration and background checks;
- Sub-Section 3.2.2 recommends training of enrollment station operators;
- Sub-Section 3.2.3 recommends use of automated means to track fingerprint image quality over time;
- Sub-Section 3.3 gives specifications for how fingerprint images are retained by agencies. [FIPS] gives requirements and options for the retention of biometric data. Retention of data supports, for example, detection of duplicate identities. When fingerprint images are retained they **shall** be stored in the format specified in sub-Section 3.3. The format specification includes the [CBEFF] header of Section 9 to implement the requirement to protect the integrity, and to allow for encryption, of the image records.
- Sub-Section 3.4 specifies the transformation of fingerprints into records suitable for transmission to the FBI for the background check.

Although FBI requirements drive the sensor specifications, the permanent electronic storage format specified in sub-Section 3.3 is an INCITS standard record and is therefore specified independently.

If an agency retains fingerprint templates, in either proprietary or standardized formats, then they shall be embedded in the [CBEFF] header of Section 9 which requires integrity protection and allows for encryption of the records.

3.2   Fingerprint image acquisition

3.2.1   Fingerprint collection

This Section specifies the capture of a full set of fingerprint images for PIV registration. A subject's fingerprints **shall** be collected according to one of the three imaging modes enumerated in Table 2.

Table 2 – Fingerprint acquisition protocols

#	Option 1 – Required presentations for plain live scan	
1	Combined plain impression of the four fingers on the right hand (no thumb)	
2	Combined plain impression of the four fingers on the left hand (no thumb)	
3	Combined impression of the two thumbs	
	Option 2 – Required presentations for rolled live scan	
1	10 separately rolled fingers	
2	Combined plain impression of the four fingers on the right hand (no thumb)	
3	Combined plain impression of the four fingers on the left hand (no thumb)	
4	Left thumb plain impression	These captures may be simultaneous (two thumbs next to each other) or sequential (one thumb at a time)
5	Right thumb plain impression	
	Option 3 – Required presentations for rolled ink on card	
1	10 separately rolled fingers	
2	Combined plain impression of the four fingers on the right hand (no thumb)	
3	Combined plain impression of the four fingers on the left hand (no thumb)	
4	Left thumb plain impression	These captures may be simultaneous (two thumbs next to each other) or sequential (one thumb at a time)
5	Right thumb plain impression	

INFORMATIVE NOTES:

1. There is no requirement that the order specified above is the order in which the images should be acquired.

2. The combined multi-finger plain-impression images are also referred to as slaps or flats. They are obtained by simultaneous placement of multiple fingers on the imaging surface without specific rolling movement.

3. Options 2 and 3 represent existing agency practice. Although Option 1 is now acceptable to the FBI, agencies may still need to implement Options 2 or 3 for transmission via the Office of Personnel Management.

For Options 1 and 2 the devices used for capture of the fingerprints **shall** have been certified by the FBI to conform to Appendix F of the FBI's Electronic Biometric Transmission Specification [EBTS, Appendix F]. For Option 3, a scan of the inked card **shall** be performed to effect conversion to electronic form. The scanner **shall** be certified by the FBI as being compliant with [EBTS, Appendix F]. The scanning is needed to produce fingerprints in the digital formats of Sub-sections 3.3 and 3.4. The FBI specifications include width and height specifications for the imaging surface. The native scanning resolution of the device **shall** be 197 pixels per centimeter (500 pixels per inch) in both the horizontal and vertical directions. These specifications comply with the FBI submission requirements and with the Image Acquisition Setting Level 31 of the Finger Image-Based Data Interchange Format standard, INCITS 381 [FINGSTD].

For live-scan acquisition i.e., options 1 and 2, the enrollment client software should display the images to the attending operator. The operator should repeat acquisition if the ridge structure is not clear, broken, or incomplete in the displayed images.

The procedure for the collection of fingerprints, presented in Table 3, **shall** be followed. The procedure **shall** employ the NIST Fingerprint Image Quality [NFIQ] algorithm[4] to initiate any needed reacquisition of the images. An attending official **shall** be present at the time of fingerprint capture. The agency **shall** employ measures to ensure the quality of acquisition and guard against faulty presentation, whether malicious or unintentional. Quality assessment might be an integral function of the acquisition device or might be implemented by the attending official. In any case, the agency **shall** ensure that the applicant does not swap finger positions or hands, occlude fingers, or misalign or misplace the fingers. Particularly, because it is common during collection of multi-finger plain impressions for little fingers (i.e., positions 05 and 10) to not be long enough to reach the imaging platen, it is accepted practice for the hand to be placed at an angle to the horizontal to ensure imaging of all four fingers. Although this is not needed with newer large-platen devices the official **shall** in all cases take care to image all fingers completely. The procedure requires segmentation of the multi-finger plain impressions; this operation may be assisted by the attending official.

Table 3 – Quality control procedure for acquisition of a full set of fingerprint images

Step	Action
1.	Attending official should start by inspecting fingers and require absence of dirt, coatings, gels, and other foreign material.
2.	Official should ensure imaging surface of the sensor or the paper card is clean.
3.	Acquire fingerprints according to Option 1, 2, or 3 in Table 2. For Option 3, scan the inked card using [EBTS, Appendix F] certified scanner.
4.	Segment the multi-finger plain impression images into single-finger images. Automated segmentation is recommended. Attending official should inspect the boundaries of the automatic segmentation and correct any failures, perhaps via an interactive graphical user interface.
5.	Compute NFIQ value[5] for thumbs and index fingers. If all have NFIQ values 1, 2, or 3 (i.e., , good quality) then go to step 8.
6.	Repeat steps 2-5 up to three more times.
7.	If after four acquisitions the index fingers and thumbs do not all have NFIQ values of 1, 2 or 3 then select that set, acquired in step 3 and segmented in step 4, for which the mean of the NFIQ values of the left index, right index, left thumb, and right thumb is minimum (i.e., of best quality). If all of the index finger and thumb quality values are unavailable (perhaps because of injury to one or more of those fingers) then use the last set from step 3 of those fingers that are available, without any application of NFIQ.
8.	Prepare and store the final records per Sub-sections 3.3 and 3.4

Ordinarily, all ten fingerprints **shall** be imaged in this process; however, if one or more fingers are not available (for instance, because of amputation) then as many fingers as are available **shall** be imaged. When fewer than ten fingers are collected, the FBI background transaction of Sub-section 3.4 requires field AMP 2.084 of the accompanying Type 2 record [see EBTS, Appendix C] to have labels indicating fingers that are amputated or otherwise not imaged.

---

[4] A second version of the NFIQ algorithm is expected 9/13. This should a) produce quality values that better predict accuracy, b) offer finer control of quality thresholds and c) offer additional capabilities. http://www.nist.gov/itl/iad/ig/development_nfiq_2.cfm

[5] Given an input image, NFIQ (version 1) returns a value from 1 (excellent) to 5 (bad).

3.2.2 Training of PIV fingerprint collection staff

Quality of the biometric data is critical to the success of a biometric application. This is particularly true for enrollment data that typically persists for years. As enrollment is an attended operation, the operator is important to the collection of high quality data. Attending staff should be trained to maintain and clean the sensor, and to collect in accordance with manufacturer's guidance and this document. Specifically Agencies **shall** apprise staff that:

— Low humidity - typical in winter – causes dry fingers from which good images are more difficult to collect. This risk can be mitigated by measurement and appropriate use of supplemental humidification. Fingers may be lightly moisturized.

— Exposure of biometric equipment to bright light sources, such as direct sunlight, is generally adverse for collection of fingerprints.

— The background check can be defeated by mutilation of the fingerprints e.g.,either temporarily (e.g.,by burns or abrasives) or permanently (e.g.,by surgical means). In addition certain medications can cause loss of fingerprint ridge structure. It is recommended that collection of fingerprints from applicants with finger injuries be delayed until the fingers heal.

3.2.3 Monitoring overall enrollment quality

In order to track enrollment quality over time, a numerical summary of operational quality may be computed as a management indicator. If computed, this summary **shall** be computed from the NFIQ values of primary fingers of all PIV Card applicants processed in each calendar month. If computed, the summary **shall** be computed using the method of NIST Interagency Report 7422 [NFIQ SUMMARY] which uses a simple formula to aggregate NFIQ values. This calculation is readily automated.

Managers can track this over time, across collection sites or stations, over different populations (e.g., contractors vs. employees), across functions (PIV issuance vs. re-issuance), or even across fingers. Managers can use aggregated quality indicators as a tool to identify fingerprint collection problems. These may be due to changes in the physical environment or unintended changes in operating procedures.

3.3 Fingerprint image format for images retained by agencies

This Section specifies a common data format record for the retention of the fingerprint images collected in Sub-section 3.2. Specifically any fingerprint images enrolled or otherwise retained by agencies **shall** be formatted according to the INCITS 381-2004 finger image based interchange format standard [FINGSTD]. This set **should** include single-finger images. These **shall** be obtained by segmentation of the plain multi-finger images gathered in accordance with Options 1, 2 or 3 of Table 2, and the single plain thumb impressions from presentations 4 & 5 of Options 2 and 3. These images **shall** be placed into a single [FINGSTD] record. The record may also include the associated multi-finger plain impressions and the rolled images. This document ([800-76]) does not specify uses for any single-finger rolled images gathered according to Options 2 or 3 of Table 2. The record **shall** be wrapped in the CBEFF structure described in Section 9. Agencies may encrypt this data per the provisions of Section 9, Table 14, Note 2.

Table 4 is a clause-by-clause profile of [FINGSTD] for PIV. Its structure is as follows.

— Rows 1-10 give normative content in INCITS 381 on the kind of images to be acquired.

— Row 11 requires the CBEFF structure of Section 9 (i.e., header of Table 14, digital signature of Section 9.3).

— Rows 12-27 give PIV specifications for the fields of the General Record Header of [FINGSTD, Table 2]. These are common to all images in the record.

— Similarly, rows 28-36 provide specifications for the Finger Image Header Record in Table 4 of [FINGSTD]. The "PIV Conformance" column provides PIV specific practice and parameter defaults of the standard.

While INCITS 381 has been revised by the INCITS M1 committee, the 2004 edition is sufficient for PIV so the 2009 revision is irrelevant to PIV; however implementations should respect the version number on Line 14 of Table 4.

To assist implementers, NIST has made [FINGSTD] sample data available[6].

Table 4 – INCITS 381 profile for agency retention of fingerprint Images

			Clause title and/or field name (Numbers in parentheses are [FINGSTD] clause numbers)	INCITS 381-2004 Field or content	Value required	PIV Conformance Values allowed	Informative Remarks
1.			Byte and bit ordering (5.1)	NC		A	Big Endian MSB then LSB
2.			Scan sequence (5.2)	NC		A	
3.			Image acquisition reqs. (6)	NC		Level 31	See [FINGSTD] and also Table 1
4.			Pixel Aspect Ratio (6.1)	NC		A	1:1
5.			Pixel Depth (6.2)	NC		A	Level 31 - 8
6.			Grayscale data (6.3)	NC		A	Level 31 - 1 byte per pixel
7.			Dynamic Range (6.4)	NC		A	Level 31 - 200 gray levels
8.			Scan resolution (6.5)	NC		A	Level 31 - 500 ppi
9.			Image resolution (6.6)	NC		197	Pixels per centimeter - no interpolation
10.			Fingerprint image location (6.7)	NC		A	Slap placement info, centering
11.			CBEFF Header (7)	MF	MV	Patron Format PIV	Multi-field CBEFF Header, Sec. 7.3
12.			General Record Header (7.1)	NC		A	
13.	Finger image record format		Format Identifier (7.1.1)	MF	MV	0x46495200	i.e., ASCII "FIR\0"
14.			Version Number (7.1.2)	MF	MV	0x30313000	i.e., ASCII "010\0"
15.			Record Length (7.1.3)	MF	MV	MIT	Size excluding CBEFF structure
16.			CBEFF Product Owner (7.1.4)	MF	MV	> 0	CBEFF PID.
17.			CBEFF Product Identifier Type (7.1.4)	MF	MV	> 0	
18.			Capture Device ID (7.1.5)	MF	MV	MIT	Vendor specified. See Note 1
19.			Image Acquisition Level (7.1.6)	MF	MV	31	Settings Level 31
20.			Number of Images (7.1.7)	MF	MV	MIT	Denote by K, see lines 28-37, see Notes 2-4
21.			Scale units (7.1.8)	MF	MV	0x02	Centimeters
22.			Scan resolution (horz) (7.1.9)	MF	MV	197	Pixels per centimeter
23.			Scan resolution (vert) (7.1.10)	MF	MV	197	
24.			Image resolution (horz) (7.1.11)	MF	MV	197	
25.			Image resolution (vert) (7.1.12)	MF	MV	197	
26.			Pixel Depth (7.1.13)	MF	MV	8	Grayscale with 256 levels
27.			Image compression algorithm (7.1.14)	MF	MV	0 or 2	Uncompressed or WSQ 3.1 See Notes 5 and 6.
28.			Reserved (7.1.15)	MF	MV	0	Two bytes, see Note 12
29.	K fingerprints, or multi-finger prints		Finger data block length (7.2.1)	MF	MV	MIT	
30.			Finger position (7.2.2)	MF	MV	MIT	
31.			Count of views (7.2.3)	MF	MV	≥ 1	M views of this finger, see Note 7
32.		M finger views	View number (7.2.4)	MF	MV	MIT	
33.			Finger image quality (7.2.5)	MF	MV	20,40,60,80,100, 254	Transformed NFIQ. See Notes 8 and 9
34.			Impression type (7.2.6)	MF	MV	0 or 2	See ANSI NIST ITL 1-2000
35.			Horizontal line length (7.2.7)	MF	MV	MIT	See Note 10
36.			Vertical line length (7.2.8)	MF	MV	MIT	
37.			Reserved (no clause)	MF	MV	0	See Note 11
38.			Finger image data (7.2.9)	MF	MV	MIT	Uncompressed or compressed WSQ Data
39.			CBEFF Signature Block	MF	MV		See Section 9.3 of this document
END OF TABLE							

Acronym		Meaning
MF	mandatory field	[FINGSTD] mandates a field **shall** be present in the record
MV	mandatory value	[FINGSTD] mandates a meaningful value for this field
NC	normative content	[FINGSTD] gives normative practice for PIV. Such clauses do not define a field in the FIR.
A	as required by standard	For PIV, value or practice is as specified in [FINGSTD]
MIT	mandatory at time of instantiation	For PIV, mandatory value that **shall** be determined at the time the record is instantiated and **shall** follow the practice specified in [FINGSTD]

NORMATIVE NOTES:

[6] Fingerprint images conformant to this specification exist at http://www.itl.nist.gov/iad/894.03/nigos/piv_sample_data.html and these were prepared using NIST software available from http://www.itl.nist.gov/iad/894.03/nigos/incits.html

1. The Capture Device ID should indicate the hardware model. The CBEFF PID [FINGSTD, 7.1.4] should indicate the firmware or software version.

2. If certain fingers cannot be imaged, the value of this field **shall** be decremented accordingly.

3. The left and right four-finger images and two-thumb images may also be included. The value of this field **shall** be incremented accordingly.

4. For PIV enrollment sets, the number of images will ordinarily be thirteen (that is, the ten segmented images from the multi-finger plain impressions, and the three plain impressions themselves, 10+4+4+2) or fourteen (if the plain thumb impressions were imaged sequentially, 10+4+4+1+1).

5. Images **shall** either be uncompressed or compressed using an implementation of the Wavelet Scalar Quantization (WSQ) algorithm that has been certified by the FBI. As of February 2011, Version 3.1 of the WSQ algorithm **shall** be used [WSQ31]. The FBI's requirement for a 15:1 nominal compression ratio **shall** apply.

6. Image compression should only be applied after the record content has been prepared, and the NFIQ quality values have been computed.

7. The term view refers to the number of images of that particular finger (position). This value would exceed one if imaging has been repeated. Inclusion of more than one image of a finger can afford some benefit in a matching process. This document recommends that any additionally available images (say, from a PIV Card re-issuance procedure) with quality value 1 to 3 should be included in the record. In all cases the images **shall** be stored in order of capture date, with newest first.

8. Quality values **shall** be present. These **shall** be calculated from the NIST Fingerprint Image Quality (NFIQ) method described in [NFIQ] using the formula $Q = 20*(6 - NFIQ)$. This scale reversal ensures that high quality values connote high predicted performance and consistency with the dictionary definition. The values are intended to be predictive of the relative accuracy of a minutia based fingerprint matching system. It is recommended that a user should be prompted to first attempt authentication using the finger with the highest quality, regardless of whether this is the primary or secondary finger.

9. The quality value **shall** be set to 254 (the [FINGSTD] code for undefined) if this record is not a single finger print (i.e., , it is a multi-finger image, or a palm print) or if the NFIQ implementation fails.

10. There is no restriction on the image size. However non-background pixels of the target finger **shall** be retained (i.e., cropping of the image data is prohibited)

11. [FINGSTD, Table 4] refers to a single-byte field labeled "reserved", but there is no corresponding clause to formally define it. The M1 committee has undertaken to resolve this by inserting a new sub-clause to require inclusion of the "Reserved" field. This will appear in a revision of [FINGSTD]. In any case, PIV implementations **shall** include the single byte field, setting the value to 0.

12. Line 27 indicates that the "Reserved" field **shall** have length 2 bytes. [FINGSTD, 7.1.15] indicates a length of 4 bytes which disagrees with the value in [FINGSTD, Table 2]. The INCITS M1 committee has indicated 2 bytes is the correct value. PIV implementations **shall** include the 2 byte field, setting the value to 0.

3.4 Fingerprint image specifications for background checks

PIV fingerprint images transmitted to the FBI as part of the background checking process **shall** be formatted according to the ANSI/NIST-ITL 1-2011 standard [AN2011] and the CJIS-RS-0010 [EBTS] specification. Such records **shall** be prepared from, and contain, only those images collected per specifications in Section 3.2.

Table 5 enumerates the appropriate transaction formats for the three acquisition options of Section 3.2. The FBI documentation [EBTS] should be consulted for definitive requirements.

Table 5 – Record types for background checks

Option	Transaction Data Format in [AN2011]	Reference
1	Three Type 14 records (see Note 1)	[EBTS, Appendix N].
2 or 3	Fourteen Type 4 records (see Notes 1 + 2)	Clause 3.1.1.4 "Federal Applicant User Fee" of [EBTS]

NORMATIVE NOTES:

1. All types of transactions with the FBI require both a Type 1 and Type 2 record to accompany the data; see [AN2011, Table 2]. The Type 2 record supports labeling of missing fingers.

2. Fourteen records, one for each of 10 fingers, one for each four-finger plain, and one for each thumb (segmented from a two-thumb image if necessary).

4.1 Scope

This Section specifies how the PIV mandatory biometric elements specified in [FIPS] are to be generated and stored. This specification applies to templates stored within the PIV Card, and to [MINUSTD] templates otherwise retained by agencies. The templates constitute the enrollment biometrics for PIV authentication and as such are supported by a high quality image acquisition specification, and an FBI-certified compression format. The specification of a standardized template here enables use of the PIV Card in a multi-vendor product environment.

4.2 Source images

Two [MINUSTD] fingerprint templates **shall** be stored on the PIV Card; these are hereafter referred to as PIV Card templates. These **shall** be prepared from images of the primary and secondary fingers. These fingers should be selected on the basis of:

— **Availability:** Ability of individuals to mechanically place the finger on a generic sensor – this contraindicates ring fingers, and sometimes thumbs.

— **Quality:** High quality fingers, preferably those of NFIQ = 1 or 2, should be used. Lower quality fingers should only be used if recapture (per Section 3.2.1) has failed and if other items in this list preclude use of better quality fingers.

— **Control:** Ability to use fine motor control in placing the finger on a sensor – for most individuals this indicates use of index fingers.

— **Handedness:** Individuals should favor their preferred hand, for most people this is the right hand.

— **Injury:** Presence of permanent or temporary injury to the friction ridge structure, or the finger itself – this contraindicates use of afflicted fingers.

— **Area:** Physical area of the finger's volar pad – this favors use of thumbs, and contraindicates little "pinky" fingers.

— **Two-finger sensors:** If two-finger sensors are deployed and used, adjacent fingers can be placed simultaneously.

— **Sensor placement:** If the fingerprint sensor is to the side of a user vs. in front (as for the driver of a vehicle), the fingers from the same hand might be used.

Thus a PIV Card applicant, in consultation with an attending operator, should select primary and secondary fingers with the following being the default, in descending order of priority.

| 1. Preferred index | 3. Preferred middle | 5. Preferred thumb | 7. Preferred ring | 9. Preferred little |
| 2. Other index | 4. Other middle | 6. Other thumb | 8. Other ring | 10. Other little |

These images **shall** be either:

— those obtained by segmenting the initial plain impressions of the full set of fingerprints captured during PIV Registration and stored in Row 8 of Table 3, or

— new images collected and matched against the initial plain impressions (see [FIPS]).

Significant rotation, exceeding 30 degrees, of the multi-finger plain impressions (for example, that which can occur when four fingers are imaged using a narrow platen) **shall** be removed prior to, or as part of, the generation of the mandatory minutiae templates. The rotation angle **shall** be that which makes the inter-phalangeal creases approximately horizontal or, equivalently, the inter-finger spaces approximately vertical. This requirement supports interoperable fingerprint matching.

4.3 Card issuance

[FIPS] establishes requirements on authentication of card applicants for example to bind the PIV Cardholder to the individual whose background was checked. This authentication **shall** use images collected using either a [EBTS, Appendix F] multi-finger fingerprint imaging device of Section 3.2, or a [SINGFING] device of Section 4.7.

4.4 Minutia record for off-card authentication

4.4.1 Use of a standard

PIV Card templates **shall** be a conformant instance of the INCITS 378-2004 [MINUSTD] minutiae template standard. A standard record is used to satisfy global interoperability objectives. Second generation standards have been published since the first PIV specification appeared in 2005; these standards are not cited in SP 800-76-2 for the reasons given in Section 1.6. Implementations **shall** therefore respect the version number on Line 14 of Table 6.

4.4.2 General case

The minutiae from both the primary and secondary fingers **shall** reside within a single INCITS 378 record. This means that there will be one instance of the "General Record Header" [MINUSTD, 6.2-6.4], and two instances of the "Finger View Record" [MINUSTD, 6.5]. The entire Table 6 record **shall** be wrapped in a single instance of the CBEFF structure specified in Section 8 prior to storage on the PIV Card. The PIV Card templates **shall** not be encrypted.

Table 6 is a profile of the generic [MINUSTD] standard. Its specifications **shall** apply to all minutiae templates placed on PIV Cards. These constraints are included to promote highly accurate and interoperable personal identity verification. This document recommends that the minutiae records should be prepared soon after the images are captured and before they are compressed for storage.

To assist implementers, NIST has made [MINUSTD] sample data available[7]. Implementers should buy [MINUSTD] for information such as the number of bits used in each field.

Table 6 – INCITS 378 profile for PIV Card templates

		Clause title and/or field name (Numbers in parentheses are [MINUSTD] clause numbers)	INCITS 378-2004 Field or content	Value Required	PIV Conformance Values Allowed	Informative Remarks
1.		Principle (5.1)	NC		A	Defines fingerprint minutiae
2.		Minutia Type (5.2)			See Note 1	[MINUSTD, 5.2] defines minutiae type but contains no normative content
3.		Minutia Location : Coordinate System (5.3.1)	NC		A	
4.		Minutia Location : Minutia Placement on a Ridge Ending (5.3.2)	NC		A	Minutia placement and angle are influential on accuracy and interoperability. Developers should ensure the listed requirements are actually achieved by their minutia detection algorithms.
5.		Minutia Location : Minutia Placement on a Ridge Bifurcation (5.3.3)	NC		A	
6.		Minutia Location : Minutia Placement on Other Minutia Types (5.3.4)	NC		See Note 1	
7.		Minutia Direction : Angle Conventions (5.4.1)	NC		A	In addition, correct detection of true minutiae, and correct suppression of false minutiae have been shown to influence interoperability [BAZIN, MANSFIELD].
8.		Minutia Direction : Angle of a Ridge Ending (5.4.2)	NC		A	
9.		Minutia Direction : Angle of a Ridge Bifurcation (5.4.3)	NC		A	
10.		Byte Ordering (6.2)	NC		A	Big Endian, unsigned integers
11.		Minutia Record Organization (6.3)	NC		A	
12.	General Record Header	CBEFF Record Header (6.4)	MF	MV	Patron format PIV	Multi-field CBEFF Header, Sec. 9.2. This wrapper is required by PIV. It is not present in [MINUSTD]
13.		Format Identifier (6.4.1)	MF	MV	0x464D5200	i.e., ASCII "FMR\0"
14.		Version Number (6.4.2)	MF	MV	0x20323000	i.e., ASCII " 20\0" which is INCITS 378-2004. See Note 2
15.		Record Length (6.4.3)	MF	MV	26 ≤ L ≤ 1574	This connotes a 2 byte field. See Note 3
16.		CBEFF Product Identifier Owner (6.4.4)	MF	MV	> 0	See Note 4
17.		CBEFF Product Identifier Type (6.4.4)	MF	MV	> 0	See Note 4
18.		Capture Equipment Compliance (6.4.5)	MF	MV	1000b	Sensor complies with EBTS, Appendix F per PIV Registration requirement
19.		Capture Equipment ID (6.4.6)	MF	MV	> 0	See Note 5
20.		Size of Scanned Image in x direction (6.4.7)	MF	MV	MIT	See Note 11
21.		Size of Scanned Image in y direction (6.4.8)	MF	MV	MIT	
22.		X (horizontal) resolution (6.4.9)	MF	MV	197	Parent images conform to Section 4.2

---

[7] Minutiae records conformant to the PIV specification are here http://www.itl.nist.gov/iad/894.03/nigos/piv_sample_data.html and these were prepared using NIST software available from http://www.itl.nist.gov/iad/894.03/nigos/incits.html

		Clause title and/or field name (Numbers in parentheses are [MINUSTD] clause numbers)	INCITS 378-2004 Field or content	Value Required	PIV Conformance Values Allowed	Informative Remarks
23.		Y (vertical) resolution (6.4.10)	MF	MV	197	
24.		Number of Finger Views (6.4.11)	MF	MV	2	Once each for primary and secondary
25.		Reserved Byte (6.4.12)	MF	MV	0	
26.		Finger View Header (6.5.1)	NC		A	
27.		Finger Position (6.5.1.1)	MF	MV	MIT	
28.		View Number (6.5.1.2)	MF	MV	0	See Note 10
29.		Impression Type (6.5.1.3)	MF	MV	0 or 2	Plain live or non-live scan images.
30.		Finger Quality (6.5.1.4)	MF	MV	20,40,60,80,100, 254, 255	See Note 6
31.		Number of Minutiae (6.5.1.5)	MF	MV	$0 \leq M \leq 128$	M minutiae data records follow. If only M $\leq$ 10 minutiae are found re-enrollment should be attempted.
32.		Minutiae Type (6.5.2.1)	MF	MV	01b, 10b, or 00b	See Note 1
33.		Minutiae Position (6.5.2.2)	MF	MV	MIT	See Note 7
34.		Minutiae Angle (6.5.2.3)	MF	MV	MIT	See Note 8
35.		Minutiae Quality (6.5.2.4)	MF	MV	MIT	This may be populated.
36.		Extended Data Block Length (6.6.1.1)	MF	MV	0	See Note 9
37.		CBEFF Signature Block	MF	MV		See Section 9.3 of this document

END OF TABLE

Acronym		Meaning
MF	mandatory field	[MINUSTD] requires a field **shall** be present in the FMR
MV	mandatory value	[MINUSTD] requires a meaningful value for a field
NC	normative content	[MINUSTD] gives normative practice for PIV. Such clauses do not define a field in the FMR.
A	as required	For PIV, value or practice is as normatively specified in [MINUSTD].
MIT	mandatory at time of instantiation	For PIV, mandatory value that **shall** be determined at the time the record is instantiated and **shall** follow the practice specified in [MINUSTD]

NORMATIVE NOTES:

1. [MINUSTD] requires that each stored minutia have a type associated with it. For PIV, the mandatory card templates **shall** contain minutiae of type ridge ending or ridge bifurcation. These types are defined in [MINUSTD, 5.3.{2,3}]. Other types of minutiae, such as trifurcations and crossovers, **shall** not be included in PIV Card templates. However, for those minutiae where it is not possible to reliably distinguish between a ridge ending and a bifurcation, the category of "other" **shall** be assigned and encoded using bit values 00b. The angle and location for a minutia of type "other" should be the angle and location that would have applied to the corresponding ridge ending or bifurcation depending on which one the encoding algorithm determines to be the most likely for that particular minutiae. This is a common characteristic of "inked" impressions that exhibit ridge endings being converted to bifurcations and vice-versa due to over- or under-inking in the image.

2. The second paragraph of [MINUSTD, 6.4.2] refers both to an ASCII space and "three ASCII numerals" mentioned in the first paragraph. The practice of using an ASCII space character as the first character of the version number **shall** be followed: " 20\0" i.e., 0x20323000.

3. The length of the entire record **shall** fit within the container size limits specified in [800-73]. These limits apply to the entire CBEFF wrapped and signed entity, not just the [MINUSTD] record.

4. Both fields ("Owner" and "Type") of the CBEFF Product Identifier of [MINUSTD, Clause 6.4.4] **shall** be non-zero. The two most significant bytes **shall** identify the vendor, and the two least significant bytes **shall** identify the version number of that supplier's minutiae detection algorithm.

5. The Capture Equipment ID **shall** be reported. Its use may improve interoperability.

6. The quality value **shall** be that computed for the parent image using [NFIQ] and reported here as Q = 20*(6 - NFIQ). A value of "255" **shall** be assigned when fingerprints are temporarily unusable for matching. A value of "254" **shall** be assigned when the fingerprints are permanently unusable.

7. All coordinates and angles for minutiae **shall** be recorded with respect to the original finger image. They **shall** not be recorded with respect to any sub-image(s) created during the template creation process.

8. Determination of the minutia direction can be extracted from each skeleton bifurcation. The three legs of every skeleton bifurcation must be examined and the endpoint of each leg determined. Figures 2A through 2C illustrate the three methods used for determining the end of a leg. The ending is established according to the event that occurs first:

- o The 32nd pixel – see Figures 2A and 2B – or
- o The end of skeleton leg if greater than 10 pixels (legs shorter are not used) – see Figure 2B – or
- o A second bifurcation is encountered before the 32nd pixel – see Figure 2C.

The angle of the minutiae is determined by constructing three virtual rays originating at the bifurcation point and extending to the end of each leg. The smallest of the three angles formed by the rays is bisected to indicate the minutiae direction.

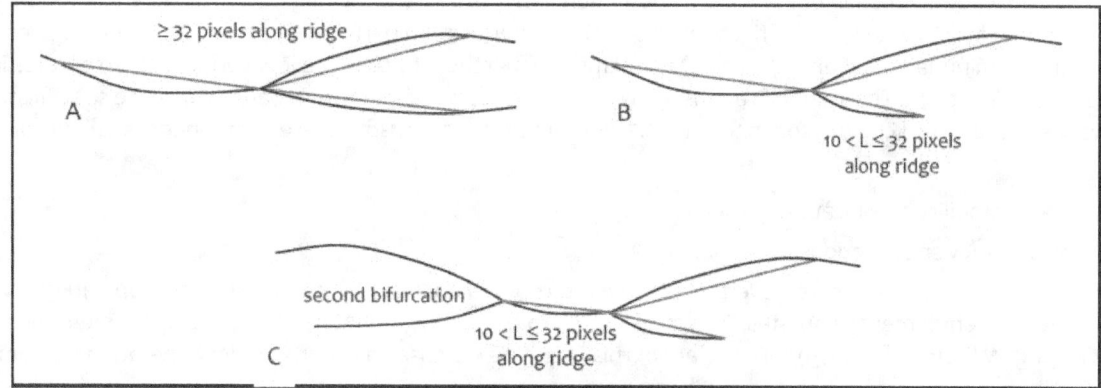

Figure 2 – Minutiae angle determination

Extensive, refined and complete guidance on minutia detection and estimation appears in INCITS 378:2009 Clause 6. That standard is the revision of INCITS 378-2004 [MINUSTD]. While PIV still requires [MINUSTD] for off-card comparison template formatting, the 2009 standard might be consulted because it gives elaborate information on the semantic (placement + selection) aspects associated with this note.

9. The mandatory value of zero codifies the PIV specification that templates **shall** not include extended data.

10. Per [MINUSTD, 6.5.1.2] this view number field **shall** have value 0 for the primary finger and 0 for the secondary finger. The combination of view number and finger position uniquely identifies each template.

11. [MINUSTD] does not specify how to report the image sizes in the header when two or more views are included in the record and these were derived from images of different sizes. For PIV, the width on Line 20 **shall** be the larger of the widths of the two input images. Similarly the height on Line 21 **shall** be the larger of the heights of the two input images.

4.4.3 Special case for individuals who cannot be fingerprinted

If two fingerprints have never been collected (e.g., because of injury, amputation, or persistent poor quality), or all fingerprint authentication attempts fail during Section 4.3 card issuance, then the PIV Card **shall** be populated with the standardized minutia record of Section 4.4 which

- has two empty views (i.e., there are zero minutiae, such that Table 6, Line 31 **shall** be zero),
- is digitally signed as usual using the properly populated CBEFF structure of Section 9,
- has fingerprint qualities (Table 6, Line 30) assigned 255 for temporarily unusable, or 254 for permanently unusable, fingerprints, and
- overrides the CBEFF quality values (Table 14, Line 11) with -1 indicating temporarily, and -2 permanently unusable fingerprints.

[FIPS] allows iris or face biometrics (see Sections 6 and 7) for some PIV operations for applicants with unavailable or unusable fingerprints.

If only one finger is available, the first view **shall** be populated and the second view **shall** be empty, as above. Authentication systems encountering cards populated with empty minutia templates might use iris authentication, if the data is present on card.

NOTE   Minutia detection and matching algorithms continue to improve. Their accuracies have been measured on reference data sets [MINEX04]. Some certified implementations are significantly more accurate than others, affording lower false match rates for equal false rejection rates.

4.5   Performance specifications for PIV compliance

4.5.1   Background and scope

The intent of the [FIPS] specification of a government-wide interoperable biometric is to support cross-vendor and cross-agency authentication of PIV Cards. These multi-party aspects cause fingerprint recognition accuracy to vary, as documented in [MINEX04]. To mitigate against poor authentication performance this document requires template generators (minutia detection algorithms) and template matchers to produce low verification error rates in interoperability tests. These specifications apply to off-card comparison of templates - separate specifications are advanced for on-card comparison in Section 5.7. For off-card comparison, these components **shall** perform according to

— interoperability specifications of sections 4.5.2, and

— the accuracy specifications of Section 4.5.3.

The criteria implement the core government-wide interoperability objectives of HSPD-12 by populating PIV Cards with interoperable enrollment templates. This is necessary to exclude systematically incorrect implementations of the underlying [MINUSTD] from PIV. The effect of this is to give increased assurance of low operational error rates.

4.5.2   Minimum interoperability specification

The core cross-vendor interoperability specification is met by establishing requirements on template generators and template matchers as described in the following two sub-sections.

4.5.2.1   Conformance of template generators

A template generator is certified on the basis of the conformance of its output, its speed of computation, and on the error rates observed when its templates are matched. A template generator **shall** be certified only if

1. it converts all input PIV representative enrollment images to Table 18 templates, and

2. all templates are syntactically conformant to the Table 18 profile of [MINUSTD], and

3. it converts 90% of PIV representative enrollment images to templates in fewer than 1.3 seconds[8] each, and

4. all certified matchers verify their output templates with FNMR less than or equal to 0.01 at a FMR of 0.01 (where this is calculated from the sum of scores from two finger comparisons e.g., left and right index fingers), and

5. the minutiae it reports have unique (x, y) values i.e., no two minutiae may share the same location. This requirement is additional to the minutia detection requirements of the [MINUSTD] and is instituted because non-uniqueness impedes some matching algorithms.

4.5.2.2   Conformance of template matchers

A template matcher is certified on the basis of its speed of computation, and on the error rates observed when it matches templates in interoperability tests. A template matcher **shall** be certified only if

1. it compares all pairs of Table 18 templates to scalar scores, and

2. it executes 90% of the Annex A.4 template matches in fewer than 0.1 seconds[8] each, and

---

[8] This specification applies to a commercial-off-the-shelf PC procured in 2005 and equipped with a 2GHz processor and 512 MB of main memory. This specification shall be adjusted by the testing organization to reflect significant changes of the computational platform.

3. it matches templates from all certified template generators, and the template generator accompanying the matcher, with FNMR less than or equal to 0.01 at an FMR of 0.01 (where this is calculated from scores from the sum of scores from two finger comparisons e.g., left and right index fingers).

4.5.3 Minimum accuracy specification

The interoperability criterion of Section 4.5.2.2[9] is designed to support low false rejection when templates can come from many sources (i.e., conformant [MINUSTD] template generators). This FMR value, however, is too high for operational application i.e., it is higher than the minimum accuracy requirements of Section 10. To support actual authentication of PIV Card templates, a template generator and matcher-pair **shall** be certified if

1. it meets all the interoperability criteria of sections 4.5.2.1 and 4.5.2.2, and

2. it matches single-finger native templates with FNMR less than or equal to 0.02 when the FMR is at or below 0.0001. The word native here means that all templates originate from one template generator and the provider of the template generation algorithm is the same as that of the comparison algorithm. Native mode operation will occur within an Agency that procures its template generation and matching equipment from the same provider.

4.5.4 Test method

The performance specifications of sections 4.5.2.1, 4.5.2.2 and referred to as Level 1 accuracy specifications. The specifications of Section 4.5.3 are referred to as Level 2 accuracy specifications. Both Level 1 and Level 2 criteria **shall** be tested as defined in Annex A.

INFORMATIVE NOTE   NIST's MINEX III program[10] implements these tests [MINEX-III] – see Section A.6.

4.6 Performance specifications for PIV operations

Off-card fingerprint authentication implementations **shall** be configured according to the specifications of Section 10.

4.7 Fingerprint capture

4.7.1 Scope

This Section gives specifications for fingerprint sensors used for capture of single finger images. These sensors **shall** not be used for collection of images for use in the background check i.e., the specifications are unrelated to those of Section 2.3 which govern ten-print enrollment. This document does not establish specifications on performance of four-finger segmentation algorithms.

4.7.2 Fingerprint acquisition specifications for flat capture sensors

Fingerprint sensors used for PIV authentication **shall** conform to the FBI's Image Quality Specifications For Single Finger Capture Devices [SINGFING]. The [SINGFING] specification establishes minimum sizes for the imaging platen and for the scanning resolution.

---

[9] The accuracy for interoperability criterion is that a matching algorithm recognizes all templates at a false match rate (FMR) less than or equal to 0.01, the false non-match rate (FNMR) is at or below 0.01.

[10] The MINEX III program replaces the original Ongoing MINEX procedure which ran 2007-2013. Implementations that were deemed compliant in the older program are automatically compliant to Sections 4.5.2.1 and 4.5.2.2 as tested in MINEX III.

5.1 Scope

[FIPS] allows (but does not require) Agencies to use on-card comparison (OCC) of fingerprint minutiae. This Section gives specifications for OCC for PIV. This specification includes enrollment data to be placed on the card, authentication data to be sent to the card, and OCC certification information. This Section also specifies the data structure for the storage of card parameters, and the procedure for preparation of OCC fingerprint minutiae templates from those for off-card comparisons.

[800-73] indicates where OCC data is stored and that this data is separate and different from the mandatory off-card comparison fingerprint templates. [800-73, Part 2] specifies the secure channel mechanisms to realize on-card comparison over the [FIPS]-specified interfaces.

5.2 Background

NIST conducted two studies to support the use of on-card biometric comparison in identity management applications.

— The Secure Biometric Match on Card[11] activity engaged commercial providers to execute fingerprint authentication over a contactless interface within a specific time limit. The study required privacy protection via secured communication protocols and integrity protection using cryptographic signatures computed from the biometric data. In addition, the card was authenticated to the reader. The activity has been published as NIST Interagency Report 7452 [SBMOC].

— The MINEX II evaluation was initiated to measure the core algorithmic speed and accuracy of fingerprint minutia matchers running on ISO/IEC 7816 smartcards. Conducted in phases, the test required card- and fingerprint matcher-provider teams to submit on-card comparison enabled cards. The latest results were reported in NIST Interagency Report 7477 [MINEX II].

The main attraction of on-card comparison has been that someone who finds a card has only a small chance of authenticating to it with fingerprints (related to the configured false match rate) – the fact that template data never leaves the card means that the attacker has no prior knowledge. See the related discussion in Section 5.4. In the off-card comparison world, the PIN entry requirement is used to mitigate this risk.

5.3 Approach to the use of standards

The PIV specification for on-card matching leverages international standards. Specifically, PIV Cards **shall**

— be prepared and used by executing the commands of ISO/IEC 7816-4:2005 [CARD-CMD] per [800-73, Part 2]

— embed the biometric data in the data structures defined in ISO/IEC 7816-11:2004 [CARD-BIO],

— use the core three-byte-per-minutia format defined in the ISO/IEC 19794-2:2011 [CARD-MIN] standard[12] (implementers may choose to prepare these from INCITS 378:2004 templates, as shown in Figure 3), and

— adopt certain defined constants from ISO/IEC 19785-3:2007.

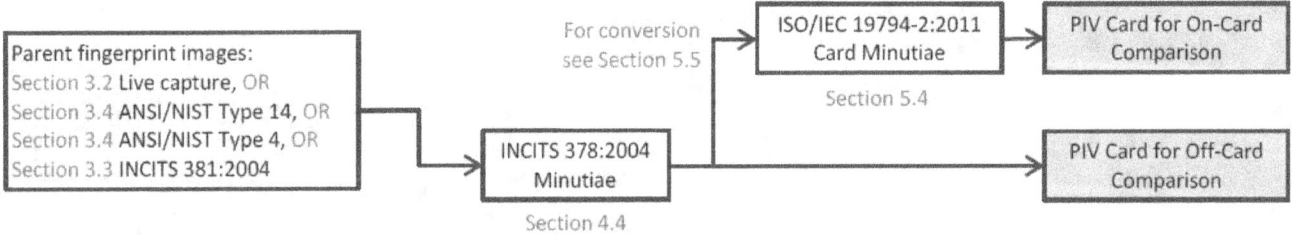

Figure 3 — Preparation of PIV Fingerprint Minutia Templates

---

[11] FIPS 201-2 uses the term "on-card biometric comparison". It is standardized and preferred over the term "match-on-card".

[12] This second edition of the minutia standard was completed on 2011-12-14.

5.4   Finger selection

This document recommends that different fingers should be imaged for off-card and on-card comparison. This mitigates an attack where off-card templates are stolen (after PIN release) and used to prepare spoofs via [REVFING] for use in on-card comparison (which is not preceded by PIN entry). This document nevertheless allows data from the same fingers in on-card and off-card comparison operations for two reasons. First is to mitigate a usability issue, namely that users might be confused as to which fingers should be presented. Second, there is always the possibility that fingerprints are stolen from other sources (including latent acquisition) and used in a spoof attack.

5.5   Data objects

5.5.1   Biometric Information Template

Each submitted card **shall** be populated with Biometric Information Templates (BITs) grouped under the BIT Group Template of Table 7 according to the requirements of [CARD-CMD, Tables 1 and 2]. The number of BITs **shall** be equal to 2, one for each finger.

Table 7 – BIT group template and profile

Tag	Len.	Value							Allowed values
7F61	Var.	BIT group template							
		Tag	Len.	Value					
		02	1	1...4 (Number of BITs in the group, corresponding to number of fingers that follow)					2
		7F60	Var.	Biometric Information Template (BIT) for the first finger					
				Tag	Len.	Value			
				83	1	Reference data qualifier used by VERIFY			'96'
				A1	Var.	Biometric Header Template (BHT) conforming to ISO/IEC 19785-3:2005			
						Tag	Len.	Value	
						81	1	Biometric type (i.e., modality, 08 = fingerprint)	08
						82	1	Biometric subtype (e.g., finger position) - These values **shall** be from ISO/IEC 19785-3:2007, NOT from [CARD-MIN].	See NOTE 2
						87	2	CBEFF BDB format owner	0101 i.e., JTC1/SC37
						88	2	0x0005 (CBEFF BDB format type)	'00 05' See NOTE 1
						B1	Var.	Biometric matching algorithm parameters. [CARD-MIN Table 14]	
								Tag	Len.	Value
								81	2	Min. and max. numbers of minutiae, see ISO/IEC 19794-2 (Sub-clause 8.3.3, Table 10)
								82	1	Minutiae order, see ISO/IEC 19794-2005 (Sub-clause 8.3.4 and Tables 11 and 12)
								83		This tag **shall** not be present Feature handling indicator, see [CARD-MIN, Table 15]
		7F60	Var.	Biometric Information Template (BIT) for the second finger					
				Tag	Len.	Value			
				83	1	Reference data qualifier used by VERIFY			'97'
				A1	Var.	Biometric Header Template (BHT) conforming to ISO/IEC 19785-3:2005			
						Tag	Len.	Value	
						81	1	Biometric type (i.e., modality, 08 = fingerprint)	08
						82	1	Biometric subtype (e.g., finger position) - These values **shall** be from ISO/IEC 19785-3:2007, NOT from [CARD-MIN].	See NOTE 2
						87	2	CBEFF BDB format owner	0101 i.e., JTC1/SC37
						88	2	0x0005 (CBEFF BDB format type)	'00 05' See NOTE 1
						B1	Var.	Biometric matching algorithm parameters. [CARD-MIN Table 14]	
								Tag	Len.	Value
								81	2	Min. and max. numbers of minutiae, see ISO/IEC 19794-2 (Sub-clause 8.3.3, Table 10)
								82	1	Minutiae order, see ISO/IEC 19794-2"2005 (Sub-clause 8.3.4 and Tables 11 and 12)
								83		This tag **shall** not be present Feature handling indicator, see [CARD-MIN, Table 15]

NORMATIVE NOTES:

1. The 0x0005 value indicates one of two definitions for reporting locations of minutiae defined in the ISO standard. This one requires that the endings of ridges be reported at the point of the valley bifurcation (versus at the ridge tip itself). These are the semantics required by INCITS 378:2004. The on-card comparison templates might reasonably be produced from the parent INCITS 378 templates.

2. Which fingers are present is encoded using integers from Table 8. The finger position codes differ in the fingerprint vs. smart-card standards. For on-card comparison data, ISO/IEC 19785-3:2007 finger position codes **shall** be used (column B). For the PIV mandatory off-card comparison templates, [MINUSTD] finger positions **shall** be used (column A). Card issuance processes shall transcode using the mapping of Table 8.

Table 8 – ISO/IEC 19794-2 and ISO/IEC 19785-3 finger position codes

Finger ID Biometric subtype	ISO/IEC 19794-2:2011 + INCITS 378:2004		ISO/IEC 19785-3:2007	
	Binary value	Hex Value	Binary value	Hex Value
	A		B	
No information given	00000b	00	00000000b	00
right thumb	00001b	01	00000101b	05
right index	00010b	02	00001001b	09
right middle	00011b	03	00001101b	0D
right ring	00100b	04	00010001b	11
right little	00101b	05	00010101b	15
left thumb	00110b	06	00000110b	06
left index	00111b	07	00001010b	0A
left middle	01000b	08	00001110b	0E
left ring	01001b	09	00010010b	12
left little	01010b	0A	00010110b	16

PIV readers involved in on-card and off-card authentication attempts **shall** heed Table 8 to correctly prompt users for which finger to present.

Note that the penultimate (i.e., FDIS) draft of ISO/IEC 19785-3:2007 erroneously set bit six to 1. The final standard and the PIV specification require that bits 6, 7 and 8 **shall** be 0.

5.5.2   Minutiae data for on card comparison

This Section defines the data to be used for on-card comparison implementations. It is included here because ISO/IEC 19794-2:2011 [CARD-MIN] and its antecedents defined multiple variants[13].

On-card comparison data in PIV **shall** conform to the ISO/IEC 19794-2:2011, Clause 9 compact on-card comparison format [CARD-MIN]. This format encodes the x and y coordinates, minutia type, and minutia angle of each minutia point in 3 bytes. This format also allows encoding of cores, deltas and ridge counts. Two cases exist:

— **Data stored on-card (i.e., enrollment data):** Neither this document nor [800-73] regulate what data must be stored on card; instead this specification only requires that the authentication instruction of [800-73, Part 2] **shall** be operable. Thus a card might validly be populated with minutiae, cores, and ancillary proprietary data.

— **Data sent to card during operations (for comparison):** This minutia data **shall** be included in the APDU specified in [800-73, Part 2] and **shall** consist of exactly 3N bytes from N minutia as specified in [CARD-MIN]. No leading tagged header information is necessary. Extended and proprietary data **shall not** be sent to the card.

The BITs of Section 5.5.1 **shall** parameterize the production of templates that a reader, or other system, sends to the PIV Card – see Section 5.6.2. This applies to both the reference templates stored on the card, and those produced during, for example, an authentication transaction.

---

[13] Particularly the ISO/IEC 19794-2:2005 standard includes three encodings (record, card-normal, card-compact), has versions with and without headers, has variants differing in their minutia placement semantics, has presence of standardized extended data (zonal quality etc.) and of non-standard, proprietary, extended data.

5.6 Preparation of the minutia templates

5.6.1 Conversion of INCITS 378 to ISO/IEC 19794-2 on-card comparison templates

Existing PIV equipment produces (CBEFF-wrapped) [MINUSTD] instances of Table 6 for off-card comparison. An OCC process may choose to use such equipment in which case [CARD-MIN] data would appropriately be prepared from live [MINUSTD] templates via the non-trivial conversion of Figure 4. The conversion operation proceeds with a pruning operation (Sec. 5.6.2.1 and 5.6.2.2), a re-encoding (conversion of 8 bit to 6 bit minutia angle, conversion from 14 bit to 8 bit position coordinates), and a sorting operation (Sec. 5.6.2.3). The order matters here due to arithmetic rounding.

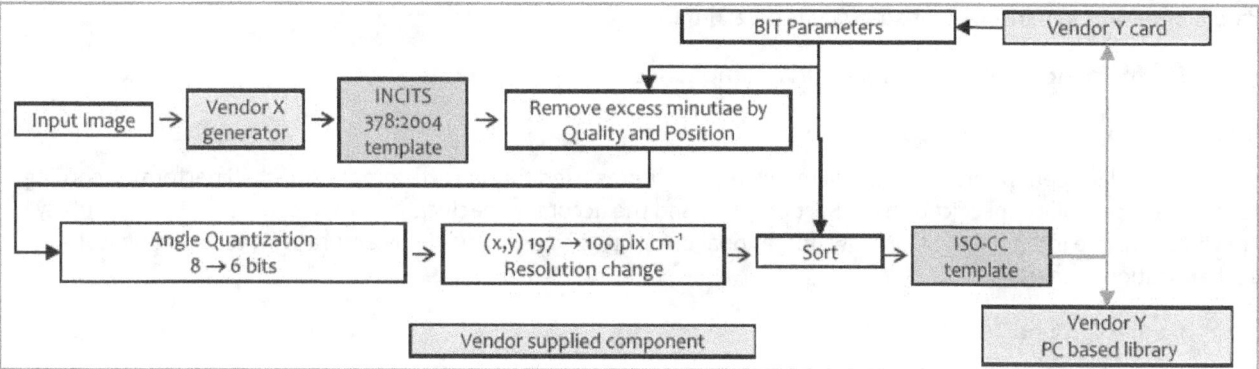

Figure 4 – Conversion of INCITS 378 to ISO/IEC 19794-2 card data

5.6.2 Effect of the BIT

5.6.2.1 Number of minutiae

The number of minutiae sent to a PIV Card for on-card comparison is not limited by this document. However, the number may be subject to limits implied by the interface specifications of [800-73, Part 2].

INFORMATIVE NOTE 1 Leading commercial minutia detectors produce a median of 41 minutiae from plain impression images with the 5% and 95% quantiles being 24 and 61 respectively over four large operational single index finger datasets.

INFORMATIVE NOTE 2 A short-length APDU command constrains the maximum number of three-byte minutiae. Implementers should consult [800-73, Part 2] on the availability of command chaining [CARD-CMD] for larger templates.

Because some templates will naturally contain 0 minutiae (i.e., the algorithm does not find any), the (off-card) client **shall** respect the minimum number indicated by the card in its BIT structure. The client **shall** either terminate the minutia-based authentication attempt or prompt for (re-)presentation of one of the enrolled fingers.

All reference and verification templates **shall** be parameterized by the BIT parameters, as follows. If,

- the value indicated in the BIT for the minimum number of minutiae is N ≥ 0,
- the value indicated in the BIT for the maximum number of minutiae is N ≤ M,
- the number of minutiae available for sending to the card for OCC is K, then
- the number of minutiae sent to the card, S, **shall** be

$$S = \begin{cases} M \text{ if} & K \geq M \\ K \text{ if} & K < M \end{cases}$$

In the case K < N, the client should initially recapture fingerprints (by re-prompting the user to replace finger on the sensor) and it that is unsuccessful should terminate the authentication.

5.6.2.2 Minutiae removal mechanism

Minutiae **shall** be removed according to the specifications of [CARD-MIN, Clause 9.3.2]. Note that because the parent [MINUSTD] template allows larger spatial extent (14 bit integers at 197 pixels cm⁻¹ off card), very large fingers may yield minutiae outside the maximum possible spatial extent that can be encoded here (8 bit integers at 100 pixels cm^{-1} on card). The pruning mechanism of Sections 5.6.2.1 and 5.6.2.2 **shall** be used to remove such minutiae.

5.6.2.3 Sort order of minutiae

The BIT associated with the on-card comparison algorithm **shall** indicate how minutiae must be sorted according to the options extended in [CARD-MIN, Clause 9.4]. However, because single finger PIV images have widths of fewer than 500 pixels when scanned at 19.7 pixels mm^{-1}, all possible minutiae coordinates **shall** be encoded in 8 bits, and the modulo sorting technique defined in [CARD-MIN] **shall not** be used.

INFORMATIVE NOTE   Open-source INCITS 378 "C" code is maintained in http://www.nist.gov/itl/iad/ig/biomdi.cfm . On-card biometric comparison client software is here: http://www.nist.gov/itl/iad/ig/bioapp.cfm.

5.7   Performance specifications for PIV compliance

5.7.1   Scope

Minutia template generators and minutia matching algorithms used for on-card comparison **shall** perform according to the interoperability specifications of Section 5.7.3, and the accuracy specifications of Section 5.7.4. The accuracy specifications are intended to afford low operational error rates by assuring highly accurate matching in typical authentication scenarios.

5.7.2   Background

NIST conducted tests of on-card comparison performance [MINEX-II]. Over four phases conducted between 2007 and 2010, the tests showed that four implementations would have attained the cross-provider interoperability specifications of Section 4.5.2.

In parallel, NIST's Secure Biometrics Match-on-Card program demonstrated cryptographic protection of the template data, and transactional durations below two seconds [SBMOC].

5.7.3   Minimum interoperability specification

The core cross-vendor interoperability specification is met by establishing requirements on paired template generators and on-card matchers as described in the following two Sub-sections.

5.7.3.1   Conformance of template generators used to prepare on-card comparison templates

Template generators **shall** conform to the specification of Section 4.5.2.1 for off-card authentication (because on-card comparison templates are generated off-card). No additional conformance specifications are defined here.

5.7.3.2   Conformance of on-card template matchers

A template matcher **shall** be certified if

1. it conforms to the off-card template matcher interoperability specifications of Section 4.5.2.2 but operating with Section 5.5.2 [CARD-MIN] format templates, and

2. it executes 90% of on-card genuine template pair comparisons (using the VERIFY command [CARD-CMD], for example) in fewer than 0.50 seconds, and

3. when implemented on a functional but modified PIV Card, and in a software library, it produces identical output similarity scores[14], and

4. it produces at least 512 unique integer scores when comparing many templates of different persons.

5.7.3.3   Test method

The performance specifications **shall** be tested according to the test defined by Annex A modified to use [CARD-MIN] templates. This test **shall** conform to the requirements of the ISO/IEC 19795-7 testing standard.

---

[14] This requirement implies non-operational requirements, namely that a prototype card shall be submitted for testing and this must allow multiple template comparisons without locking and must report similarity scores to a dedicated test laboratory application.

INFORMATIVE NOTE   NIST's MINEX IV program[15] implements this test [MINEX-IV] – see Annex A.6.

5.7.4   Minimum accuracy specification

5.7.4.1   Specification

To support operational authentication of PIV Card templates against live samples, a template generator and matcher-pair **shall** be certified if

1.   it meets all the interoperability criteria of sections 4.5.2.1 and 4.5.2.2, and

2.   it matches single-finger native templates with FNMR less than or equal to 0.02 when the FMR is at or below 0.0001. The word native here means that all templates originate from one template generator and the provider of the template generation algorithm is the same as that of the comparison algorithm. Native mode operation will occur within an Agency that procures its template generation and matching equipment from the same provider.

5.7.4.2   Test method

The performance specifications **shall** be tested according to the test defined by Annex A.

INFORMATIVE NOTE   NIST's MINEX IV program implements this test [MINEX-IV] – see Section A.6.

5.7.5   Performance specifications for PIV operations

On-card comparison authentication implementations **shall** be configured according to the specifications of Section 10.

5.8   Fingerprint capture

On-card comparison **shall** be implemented using the fingerprint sensors specified in Section 4.7.

5.9   On-card comparison interface

[FIPS] establishes requirements on interfaces to OCC implementations.

---

[15] The MINEX IV program replaces the original MINEX II proof-of-concept evaluation which ran 2007-2011.

6.1 Scope

This Section standardizes specifications for use of iris images as allowed by [FIPS]. The Section includes specifications

- for iris images stored on and off PIV Cards,
- for iris capture devices, and
- for components involved in automated recognition of PIV iris imagery.

The specifications extend the format requirements of ISO/IEC 19794-6:2011 with image quality related properties. The capture device specifications concern imaging properties of the iris camera, and software interfaces around it. The recognition component is specified in terms of minimum authentication accuracy and processing speed.

This document makes no mention of an iris template. In iris recognition, templates are proprietary non-standardized mathematical encodings[16] of information extracted from the formally standardized images that are defined in this document. Templates are not interoperable. Interoperability is achieved with standardized images. Agencies electing to retain only templates are vulnerable to supplier lock-in, and an inability to benefit from technology updates.

6.2 Background

Iris recognition affords highly accurate recognition of individuals. It has been used both for 1:1 verification and 1:N identification [UKIRIS, IREX-III] and has proven stable [NEXUS, IREX-VI] for over a decade. Moreover, iris images can be compressed to achieve small sizes [IREX-I, IREX-IV] affording rapid transmission across band-limited networks and storage on identity credentials. This aspect is leveraged below.

Digital representations of rectilinear images of the human iris have been formally standardized as ISO/IEC 19794-6:2011 [IRISSTD]. This standard, which replaces earlier editions, is a necessary component in an interoperable marketplace of iris cameras and iris recognition algorithms. The standard is used because it includes specialized image formats that support compact storage[17] on ISO/IEC 7816 Integrated Circuit cards. The PIV formats are shown in Figure 5.

Label	A	B
Example Image		
ISO/IEC 19794-6:2011	Image Type 2	Image Type 7
Properties	Parent image, typically the output of a camera of size 640x480 pixels, not necessarily centered, but conformant to Image Type 2 of [IRISSTD]. Images of this kind are not intended to be heavily compressed.	Cropped, masked and centered iris conformant to Image Type 7 of [IRISSTD]. Images of this kind can be compressed to a few kilobytes. For PIV, the recommended size is 3KB. The eyelids and sclera **shall** be masked.
PIV Role	Image captured from camera. This format is suitable for retention of iris images e.g.,in the [FIPS] chain-of-trust.	Prepared from (A), it **shall** be used if an agency chooses to store iris images on the PIV Card.

Figure 5 – Image formats of ISO/IEC 19794-6:2011

---

[16] Some (commercial) template representations are actually larger than the specialized ISO/IEC 19794-6:2011 Image Type 7 PIV Card *images* specified in this document.

[17] First generation iris image standards included a polar-coordinate encoding. This supported compact sizes but was removed from second generation standards because interoperability is sensitive to correct determination of the iris and pupil centers. A replacement format (Figure 5B) has been shown to offer accurate recognition [IREX-IV] and broad industry support [IREX-I]. It requires localization of the boundaries and the iris center. These tasks are non-trivial and are supported by quantitative tests.

6.3 Iris image specification for PIV Cards

6.3.1 General case

Iris images on PIV Cards **shall** conform to the requirements expressed in the Table 9 profile of the ISO/IEC 19794-6:2011 standard. Where required values and practice are not stated, the underlying requirements of the base standard **shall** apply. The profile defines a standard record that contains one or two specialized iris images each of size around 3 kilobytes. These images **shall** follow the semantic requirements of Image Type 7 images defined in the standard. The objective of these specifications is to afford maximum possible iris accuracy, low storage requirements, and corresponding fast read times. These requirements include centering and masking of the eyelid and sclera regions (an example is shown in Figure 5, column B). The masked regions can be very efficiently compressed. This affords small record sizes and, vitally, preservation of the iris texture.

Table 9 – ISO/IEC 19794-6 profile for iris images stored on PIV Cards

		Clause or field of ISO/IEC 19794-6	ISO/IEC 19794-6		PIV Conformance	Remarks
			Field	Value	Values Allowed	
1.		CBEFF Header	MF	MV	Patron format PIV	Multi-field CBEFF Header. Sec. 9.2.
2.		Format identifier	MF	MV	0x49495200	IIR\0 Four byte format identifier including null terminator.
3.		Version number	MF	MV	0x30323000	020\0 Second 19794-6 version - not the 2005 standard
4.	Iris General Header	Length of record	MF	MV	See NOTE 1	The length (in bytes) of the entire iris image data.
5.		Number of iris representations	MF	MV	1 or 2	Number of iris representations that follow. This **value would** ordinarily be 1. See NOTE 4.
6.		Certification flag	MF	MV	0x00	
7.		Number of eyes represented	MF	MV	1 or 2	2 if left and right are known present, else 1 if left or right is known present. If camera does not estimate eye label automatically, these **shall** be manually assigned.
Representation 1: Data for the first eye image follows						
8.		Representation Length	MF	MV		Bytes for this representation including the header + image
9.		Capture date and time	MF	MV	2011 onwards.	Capture start time in UTC
10.		Capture device technology identifier	MF	MV		
11.		Capture device vendor ID	MF	MV		Manufacturer ID
12.		Capture device type ID	MF	MV		Vendor assigned make model product ID.
13.		Quality block	MF	OIT		
14.		Representation number	MF	MIT	1 and then, optionally, 2	Representation sequence number
15.		Eye label	MF	MIT	1 or 2	Left, right. If camera does not estimate eye label automatically, these **shall** be manually assigned.
16.	Representation Header + Image Data	Image type	MF	MV	7	IMAGE_TYPE_CROPPED_AND_MASKED = 7 (07Hex) i.e., a cropped and region-of-interest masked, centered, iris image with (0,6R 0,2R) margins. See NOTE 2
17.		Image format	MF	MV	10 = 0x0A	Compression algorithm and encoding **shall** be mono JPEG 2000. The format **shall** not be PNG, RAW, or JPEG.
18.		Iris image properties bit field	MF	MIT MIT MV MV	Bits 1-2: 01 or 10 Bits 3-4: 01 or 10 Bits 5-6: 01 Bits 7-8: 01 Bit 1 is the least signif. bit. Bit 8 is the most signif. bit.	Horizontal + vertical orientation **shall** not be undefined Scan type **shall** be progressive. Compression history **shall** be none; i.e., the cropped and masked image **shall** be prepared from an uncompressed parent image.
19.		Image width, W	MF	MIT	288 ≤ W ≤ 448	Dimensions ranges, in pixels, are implied by the exact
20.		Image height, H	MF	MIT	216 ≤ H ≤ 336	[IRISSTD] margin requirements based on iris size.
21.		Bit depth	MF	MV	8	Bit depth in bits per pixel. This **shall** not be used to indicate compression level
22.		Range	MF	OIT		Required field; optionally populated.
23.		Roll angle of eye	MF	OIT	≤ 20	Camera or software should estimate roll angle. Rotation
24.		Roll angle uncertainty	MF	OIT	≤ 5	should only be applied if angle is > 20 deg.
25.		Iris centre, lowest X	MF	MV	W/2 for W odd, else	These values are redundant for Image type = 7 for which
26.		iris centre, highest X	MF	MV	W/2+1 for W even	image **shall** be exactly centered. The iris center **shall** be
27.		Iris centre, lowest Y	MF	MV	H/2 for H odd, else	estimated by the iris localization code, or if necessary by a
28.		Iris centre, highest Y	MF	MV	H/2+1 for H even	human inspector.
29.		Iris diameter, lowest	MF	MIT	D ≥ 160	These two fields are used to express a normative PIV
30.		Iris diameter, highest	MF	MIT	D ≤ 280	requirement on iris size. See NOTE 3
31.		Image length	MF		Approx 3KB for single iris,	Size of the JPEG 2000 encoded image data, in bytes, is limited

					and 6KB if two irises are stored.	by container defined in NIST Special Pub 800-73, and the size of its CBEFF header and digital signature.
Representation 2: Data for the second eye image follows						
Analogous to Representation 1, above.						
32.	CBEFF Signature Block	MF	MV			See Section 9.3 of this document

Acronym		Meaning
MF	mandatory field	[IRISSTD] requires a field **shall** be present in the IIR
MV	mandatory value	[IRISSTD] requires a meaningful value for a field
OV	optional value	[IRISSTD] allows a meaningful value or allows 0 to be used to connote "unspecified"
MIT	mandatory at time of instantiation	For PIV, mandatory value that **shall** be determined at the time the record is instantiated and **shall** follow the practice specified in [IRISSTD]
OIT	optional at time of instantiation	For PIV, optional header value that may be determined at the time the record is instantiated

NORMATIVE NOTES:

1. The entire record length plus the CBEFF header and CBEFF signature block length must be less than or equal to size specified in NIST Special Publication 800-73-4. A single image of size 3K, or two images of each of size 3K, will fit in this container. These sizes refer to the JPEG 2000 compressed iris image. JPEG 2000 implementations **shall** be executed with a bit rate input value that corresponds to a 3Kilobyte target result. Higher bit rates and higher sizes are allowed.

2. The specification of a Type 7 image requires that the image captured from the camera has sufficient margin around the iris to support the strict (0.6R, 0.2R) margin requirements of Image Type 7. During enrollment, client capture software might usefully display the result with a prototypical overlay.

3. If any captured iris has diameter outside of the range [160,280] pixels, see Section 6.7.1.3.

4. The record should include a single iris. This recommendation exists because one iris can easily satisfy 1:1 comparison accuracy objectives [IREX-III] and, moreover, a single eye will be read faster and its digital signature can be accessed and verified faster. Two eyes will be useful if one of the images is somehow of poor quality, or if one eye is somehow occasionally unavailable for authentication. Quality control of the PIV Card imagery is imperative.

6.3.2   Special case for individuals whose eyes cannot be captured

In cases where an Agency elects to place iris records on PIV Cards, there may be a few instances where iris images cannot be captured (for example, due to rare medical conditions). In these cases the Card should nevertheless be populated with the special "null" iris record of Table 10. This defines an empty but conformant instance of [IRISSTD] with an image of width and height equal to 1 pixel. The reason for doing this is to ensure that a cryptographically signed record exists on the PIV Card where it should. Implementations **shall** check the digital signature and **shall** reject all iris-based attempts at authentication

Table 10 – ISO/IEC 19794-6 profile for a null iris image stored on PIV Cards

		Clause or field of ISO/IEC 19794-6	ISO/IEC 19794-6		PIV Conformance	Remarks
			Field	Value	Values Allowed	
1.		CBEFF Header	MF	MV	Patron format PIV	Multi-field CBEFF Header. Sec. 9.2.
2.	Iris General Header	Format identifier	MF	MV	0x49495200	IIR\0 Four byte format identifier including null terminator.
3.		Version number	MF	MV	0x30323000	020\0 Second 19794-6 version - not the 2005 standard
4.		Length of record	MF	MV	See NOTE 1	The length (in bytes) of the entire iris image data.
5.		Number of iris representations	MF	MV	1	Number of iris representations that follow.
6.		Certification flag	MF	MV	0x00	
7.		Number of eyes represented	MF	MV	0	0 indicates laterality is unknown
Representation 1: Data for the only eye image follows						
8.	Representatio	Representation Length	MF	MV		Bytes for this representation including the header + image
9.		Capture date and time	MF	MV	2011 onwards.	Capture start time in UTC
10.		Capture device technology identifier	MF	MV		
11.		Capture device vendor ID	MF	MV		Manufacturer ID
12.		Capture device type ID	MF	MV		Vendor assigned make model product ID.

13.	Quality block	MF	MV	0	Zero indicates no quality blocks follow
14.	Representation number	MF	MIT	1	Representation sequence number
15.	Eye label	MF	MIT	0	Undefined which eye
16.	Image type	MF	MV	7	IMAGE_TYPE_CROPPED_AND_MASKED = 7 (07$_{Hex}$) i.e., a cropped and region-of-interest masked, centered, iris image with (0,6R 0,2R) margins. See NOTE 2
17.	Image format	MF	MV	2 = 0x2	Encoding **shall** be RAW
18.	Iris image properties bit field	MF	MV	01010101	Bit field values.
19.	Image width, W	MF	MIT	1	Dimensions ranges, in pixels, are implied by the exact [IRISSTD] margin requirements based on iris size.
20.	Image height, H	MF	MIT	1	
21.	Bit depth	MF	MV	8	Bit depth in bits per pixel. This **shall** not be used to indicate compression level
22.	Range	MF	OIT	0	Required field; optionally populated.
23.	Roll angle of eye	MF	OIT	0	Camera or software should estimate roll angle. Rotation should only be applied if angle is > 20 deg.
24.	Roll angle uncertainty	MF	OIT	0	
25.	Iris centre, lowest X	MF	MV	0	These values are redundant for Image type = 7 for which image **shall** be exactly centered. The iris center **shall** be estimated by the iris localization code, or if necessary by a human inspector.
26.	iris centre, highest X	MF	MV		
27.	Iris centre, lowest Y	MF	MV	0	
28.	Iris centre, highest Y	MF	MV		
29.	Iris diameter, lowest	MF	MIT	0	These two fields are used to express a normative PIV requirement on iris size. See NOTE 3
30.	Iris diameter, highest	MF	MIT	0	
31.	Image length	MF		1	1 byte
32.	CBEFF Signature Block	MF	MV		See Section 9.3 of this document

6.4 Iris image specification for iris images retained outside the PIV Card

This document neither requires nor precludes agencies from retaining iris images. [FIPS] recommends use of iris imagery (and face in limited cases) in cases where fingerprints cannot be captured satisfactorily. In addition, [FIPS] allows iris whenever fingerprints are used, and indicates that iris image data may be available outside the PIV Card for authentication during PIV Card issuance, re-issuance, and verification data reset transactions. If agencies elect to retain images, they **shall** be stored in the format specified in this clause. This clause establishes a profile of ISO/IEC 19794-6:2011 suited for retention of iris images outside the PIV Card. The format specification includes the [CBEFF] header of Section 9, and this requires integrity protection and allows for encryption of the image records.

Retention of data supports, for example, detection of duplicate identities.

Table 11 – ISO/IEC 19794-6 profile for iris images stored outside PIV Cards

		Clause or field of ISO/IEC 19794-6	ISO/IEC 19794-6		PIV Conformance	Remarks
			Field	Value	Values Allowed	
1.		CBEFF Header (5.3)	MF	MV	Patron format PIV	Multi-field CBEFF Header. Sec. 8.
2.		Format identifier	MF	MV	0x49495200	IIR\0 Four byte format identifier including null terminator.
3.	Iris General Header	Version number	MF	MV	0x30323000	020\0 Second 19794-6 version - not the 2005 standard
4.		Length of record	MF	MV		The length (in bytes) of the entire iris image data.
5.		Number of iris representations	MF	MV	1 or 2	Number of iris representations that follow. One iris is ample for verification tasks.
6.		Certification flag	MF	MV	0x00	Is certification information present in the representation headers?
7.		Number of eyes represented	MF	MV	1 or 2	2 if left and right are known present, else 1 if left or right is known present.
Representation 1: Data for the first eye image follows						
8.		Representation Length	MF	MV		Bytes for this representation including the header + image
9.		Capture date and time	MF	MV	2011 onwards.	Capture start time in UTC
10.		Capture device technology identifier	MF	MV	0x00 0x01	Unknown or Unspecified CMOS/CCD
11.	Representation Header + Image	Capture device vendor ID	MF	MV		Manufacturer ID
12.		Capture device type ID	MF	MV		Vendor assigned make model product ID.
13.		Quality block	MF	OIT		
14.		Representation number	MF	MIT	1 and then 2	Representation sequence number
15.		Eye label	MF	MIT	1 or 2	Left, right. If camera does not estimate eye label automatically, these **shall** be manually assigned.
16.		Image type	MF	MV	2	IMAGE_TYPE_VGA = 0x02 i.e., 640 x 480 pixels. See [IRISSTD]
17.		Image format	MF	MV	14 = 0x0E	Compression and encoding **shall** be PNG or RAW.

#	Field				
				2 = 0x02	
18.	Iris image properties bit field	MF	MIT MIT MV MV	Bits 1-2: 01 or 10 Bits 3-4: 01 or 10 Bits 5-6: 01 Bits 7-8: 01 Bit 1 is the least signif. bit. Bit 8 is the most signif. bit.	Horizontal + vertical orientation **shall** not be undefined Scan type **shall** be progressive. Compression history **shall** be none
19.	Image width, W	MF	MIT	640	width in pixels, W
20.	Image height, H	MF	MIT	480	height in pixels, H
21.	Bit depth	MF	MV	8	Bit depth in bits per pixel. This **shall** not be used to indicate compression level
22.	Range	MF	OIT		Required field; optionally populated.
23.	Roll angle of eye	MF	OIT	≤ 20	Camera or software should estimate roll angle. Rotation should only be applied if angle is > 20 deg.
24.	Roll angle uncertainty	MF	OIT	≤ 5	
25.	Iris centre, lowest X	MF	MIT		Iris need not be centered for Image type 2 but iris centre must be in a range such that margin requirements of [IRISSTD] are met.
26.	iris centre, highest X	MF	MIT		
27.	Iris centre, lowest Y	MF	MIT		
28.	Iris centre, highest Y	MF	MIT		
29.	Iris diameter, lowest	MF	MIT	≥ 160	These two fields are used to express a normative PIV requirement that iris diameter **shall** be no smaller than 160 pixels, and no larger than 280 pixels. See NOTE 1
30.	Iris diameter, highest	MF	MIT	≤ 280	
31.	Image length	MF	MIT		Size of the PNG encoded image data, in bytes, is unlimited
Representation 2: Data for the second eye image follows					
Analogous to Representation 1, above.					
32.	CBEFF Signature Block	MF	MV		See Section 9.3 of this document

NORMATIVE NOTE:

1. If any captured iris has diameter outside of the range [160,280] pixels, see Section 6.7.1.3.

6.5   Conformance of ISO/IEC 19794-6:2011 records

For the standard records of sections 6.3 and 6.4, implementers may wish to download the NIST-developed conformance test suites [BIOCTS] which include provision for testing of the ISO/IEC 19794-6:2011 PIV records. The source code for these CTSs is also available for download. The software for testing the syntactic correctness of these records exists in two forms: One runs under a conformance testing architecture; the other can be compiled and run in standalone applications. The former can test single-instances or been run in batch mode.

6.6   Iris image quality control

Agencies electing to store iris on PIV Cards should require PIV Applicants to:

1.   Remove eyeglasses, hard contact lenses, or patterned contact lenses during initial enrollment;

2.   Perform a one-to-one verification of a newly captured iris image with the image that is, or will be stored, on the Card. If this authentication fails, the client software **shall** recapture an image and repeat the matching procedure. The camera and associated software might collect several images and cross match them.

Additionally, agencies might train their operators in how to collect iris images: NIST has made available several documentary materials to support proficient collection and review of iris enrollments [IREX-V]. In any case the attending operator should:

—   Instruct the PIV Cardholder to open their eyes widely, remain still and look into the camera as designed;

—   Inspect captured images to verify that the eyes are open, not blurred, looking toward the camera, and that the iris is centered.

Automated quality control software is available also [IREX-II, IRISQUAL].

6.7   Performance specifications for PIV compliance

The core cross-provider interoperability specification is supported by establishing requirements on components preparing and matching [IRISSTD] records as described in sub-sections 6.7.1 through 6.7.3.

6.7.1 Properties of iris cameras

6.7.1.1 Scope

The following sub-sections support interoperable recognition by specifying iris camera and iris image properties. Additionally, when selecting iris cameras, Agencies should consult informative Annex B.

6.7.1.2 Format

The camera **shall** produce, possibly in conjunction with client-side software, conformant Table 11 [IRISSTD, Image Type 2] instances (suitable for use in an authentication transaction).

6.7.1.3 Iris size

All iris images prepared in PIV (for storage on cards, for authentication and other purposes) **shall** have an iris diameter between 160 and 280 pixels. If the camera or client software detects an iris of radius outside this range, re-capture of the PIV cardholder's iris images should be attempted at least two times. The recapture requirement is intended to correct out-of-focus irises that have incorrect diameter.

Interpolation of iris images to increase size **shall** not be performed, unless the physical iris size is actually below 9mm. Thus the optical design of the camera **shall** ensure that an iris of physical dimension 10 mm produces an iris of diameter at least 160 pixels in the digital image.

6.7.1.4 Rectilinear imaging and aspect ratio

The output of the camera **shall** be a rectilinear image of the iris region. The digital representation of the iris **shall** exhibit minimal projective distortion such that the vertical and horizontal scale factors are uniform to within ±2% throughout the image.

6.7.1.5 Spectral properties of the illuminant

The iris camera **shall** use dedicated infrared illuminators emitting light in the [700,900] wavelength interval. Cameras that are primarily sensitive to visible light (e.g.,as used for face photography) are not suitable for PIV and **shall** not be used. See Annex B.

6.7.1.6 Safety of the illuminant

The camera **shall** conform to relevant (irradiance and exposure duration) limits specified for infrared illumination given in [ICNIRP-LED, ICNIRP-BB] and the threshold limit values specified in [IECLAMP].

6.7.2 Specifications for iris record generators

Production of the standard PIV Card records of Section 6.3 is a non-trivial task because it requires iris detection, and localization, and preparation of the Figure 5B image. A standard record generator **shall** be certified if all of the following hold.

— It converts all PIV-representative[18] captured images to syntactically conformant Table 9 [IRISSTD, Image Type 7] instances (suitable for enrollment on PIV Cards).

— The median time taken to convert PIV-representative captured images to Table 9 [IRISSTD] records is below 0.5 seconds[19] each.

— At least one matcher verifies its uncompressed single-eye Table 9 records with false non-match rate (FNMR) no higher than the FNMR measured for the parent Table 11 images, when the threshold is set to achieve a false match rate (FMR) at or below 0.0001. This specification ensures that the crop, mask and centering operations alone do not degrade accuracy (before compression is applied).

— The images meet the [IRISSTD] requirements for cropping, centering, masking, and boundary blurring. Particularly, the eyelids and sclera **shall** be masked.

---

[18] These are 640x480 images conforming to Image Type 2 of [IRISSTD].

[19] This specification applies to a commercial-off-the-shelf PC procured in 2010 and equipped with a 2GHz processor and 8GB of main memory. This specification **shall** be adjusted by the testing organization to reflect significant changes of the computational platform.

6.7.3 Specifications for iris image matchers

A recognition algorithm is certified on the basis of its speed of computation, and on the error rates observed when it matches single-eye records. Specifically, a recognition algorithm **shall** be certified if the following hold.

- The median time taken to execute comparisons of genuine template pairs is below 0.05 seconds.
- It matches both compressed Table 9 and Table 11 [IRISSTD] records from all certified record generators with FNMR less than or equal to 0.01 at a FMR no larger than 0.0001.

6.7.4 Test methods

The performance specifications of sections 6.7.2 and 6.7.3 **shall** be tested in an offline test using sequestered image data. That test **shall** include visual inspection of the images produced for, and embedded in, the Table 9 records [IRISSTD, Image Type 7].

INFORMATIVE NOTE   NIST's IREX VIII activity is a program that implements this test.

6.8 Performance specifications for PIV operations

Iris authentication implementations **shall** be configured according to the specifications of Section 10.

7.1   Scope

[FIPS] establishes requirements and options for agency-collection, storage, and use of a facial image from PIV applicants. The facial imagery **shall** be stored in the format specified here. The face specification has a very similar format, and is functionally identical to, the ISO/IEC 19794-5:2005 face image adopted by the International Civil Aviation Organization for e-Passports [ICAO]. However, note that two images are involved in one-to-one applications:

— The enrollment image i.e., the PIV image as specified here.

— The live authentication image for which additional capture specifications are typically necessary to address subject height variations and the illumination environment (see [BSI-FACE], for example).

The image is suitable for automated face recognition and therefore fielded implementations **shall** conform to the accuracy specifications given in this Section.

7.2   Acquisition and format

This Section provides specifications for the retention of facial images. Facial images collected during PIV Registration **shall** be formatted such that they conform to INCITS 385-2004 [FACESTD]. In addition to establishing a format, [FACESTD] specifies how a face image should be acquired. This is done to improve image quality and, ultimately, performance. The images **shall** be embedded within the CBEFF structure defined in Section 9. Because [FACESTD] is generic across applications it includes either-or requirements. Table 12 is an application profile of [FACESTD] tailored for PIV. It gives concrete specifications for much of the generic content. Column 3 references [FACESTD] and columns 4 and 5 give [FACESTD] requirements. For PIV, column 6 of Table 12 gives normative practice or value specifications. The table is not conformant with the Implementation Conformance Statement [ICS] standard. Particularly, it extends the function of ICS but, because it has the needed rows, it may be useful in construction of a traditional ICS. Nevertheless, a "values supported column" [ICS, Clause 9.1] should be added by implementers for checking conformance to the specifications.

While the INCITS 385 standard has been superseded by the ISO/IEC 19794-5 standard, it is retained here because PIV has used it since inception – see Section 1.6.

Table 12 – INCITS 385 profile for PIV facial images

		Clause title and/or field name (Numbers in parentheses are [FACESTD] clause numbers)	INCITS 385-2004		PIV Conformance	Informative Remarks
			Field or content	Value Required	Values Allowed	
1.		Byte Ordering (5.2.1)	NC		A	Big Endian
2.		Numeric Values (5.2.2)	NC		A	Unsigned Integers
3.	CBEFF	CBEFF Header (5.3)	MF	MV	Patron format PIV	Multi-field CBEFF Header. Sec. 7.3.
4.		Format Identifier (5.4.1)	MF	MV	0x46414300	i.e., ASCII "FAC\0"
5.		Version Number (5.4.2)	MF	MV	0x30313000	i.e., ASCII "010\0"
6.	Facial Header	Record Length (5.4.3)	MF	MV	MIT	See Note 1
7.		Number of Facial Images (5.4.4)	MF	MV	≥ 1	One or more images (K ≥ 1). See Notes 2 and 3, and also line 20.
8.		Facial image Block Length (5.5.1)	MF	MV	MIT	
9.		Number of Feature Points (5.5.2)	MF	MV	≥ 0	Positive, if features computed
10.	Facial Info. Single instance of subject-specific info.	Gender (5.5.3)	MF	OV	OIT	These fields populated with meaningful values at agency discretion, otherwise 0 for unspecified.
11.		Eye color (5.5.4)	MF	OV	OIT	
12.		Hair color (5.5.5)	MF	OV	OIT	
13.		Feature Mask (5.5.6)	MF	OV	OIT	
14.		Expression (5.5.7)	MF	OV	1	Neutral
15.		Pose Angles (5.5.8)	MF	OV	0	Unspecified = Frontal
16.		Pose Angle Uncertainty (5.5.9)	MF	OV	0	Attended operation so should be frontal.
17.	Features	MPEG4 Features (5.6.1)	NC		OIT	
18.		Center of Facial Features (5.6.2)	NC		OIT	
19.		The Facial Feature Block Encoding (5.6.3)	OF	OV	OIT	
20.	Image Info. Each instance has	Facial Image Type (5.7.1)	MF	MV	1	See Note 4.
21.		Image Data Type (5.7.2)	MF	MV	0 or 1	See Note 5. Compression algorithm.

#	Group	Clause title and/or field name (Numbers in parentheses are [FACESTD] clause numbers)		INCITS 385-2004 Field or content	INCITS 385-2004 Value Required	PIV Conformance Values Allowed	Informative Remarks
22.	image-specific info.	Width (5.7.3)		MF	MV	MIT	See Note 7.
23.		Height (5.7.4)		MF	MV	MIT	
24.		Image Color Space (5.7.5)		MF	MV	1	sRGB. See Note 8.
25.		Source Type (5.7.6)		MF	MV	2 or 6	Digital still or digital video
26.		Device Type (vendor supplied device ID) (5.7.7)		MF	MV	MIT	
27.		Quality (5.7.8)		MF	0-100	A	[FACESTD] requires 0 (unspecified) but allowed here.
28.	Image Data	Data Structure (5.8.1)		MF	MV	MIT	Compressed Data
29.	Basic (Clause 6)	Inheritance	Inheritance (6.1)	NC		A	
30.			Image Data Encoding (6.2)	NC		A	See Note 5
31.			Image Data Compression (6.3)	NC		A	See Notes 5+6
32.		Format	Facial Header (6.4.1)	NC		A	Include 4 fields
33.			Facial Information (6.4.2)	NC		A	Include 9 fields
34.			Image Information (6.4.3)	NC		A	Include 8 fields
35.	Frontal (Clause 7)	Inheritance	Inheritance (7.1)	NC		A	Inherits Basic
36.		Scene	Purpose (7.2.1)	NC		A	frontal Annex A
37.			Pose (7.2.2)	NC		Frontal	+/- 5 degrees
38.			Expression (7.2.3)	NC		Neutral	
39.			Assistance in positioning face (7.2.4)	NC		A	Only the subject appears
40.			Shoulders (7.2.5)	NC		A	Body + Face toward camera
41.			Backgrounds (7.2.6)	NC		Annex A.4.3	Uniform
42.			Subject and scene lighting (7.2.7)	NC		A	Uniform
43.			Shadows over the face (7.2.8)	NC		A	None
44.			Eye socket shadows (7.2.9)	NC		A	None
45.			Hot spots (7.2.10)	NC		A	Should be absent. Diffuse light.
46.			Eye glasses (7.2.11)	NC		A	Subject's normal condition
47.			Eye patches (7.2.12)	NC		A	Medical only
48.		Photographic	Exposure (7.3.2)	NC		A	No saturation
49.			Focus and Depth of Field (7.3.3)	NC		A	In focus
50.			Unnatural Color (7.3.4)	NC		A	White balance
51.			Color or grayscale enhancement (7.3.5)	NC		A + no recompress	No post-processing
52.			Radial Distortion of the camera lens (7.3.6)	NC		A + Follow Annex A.8	
53.		Digital	Geometry	aspect ratio (7.4.2.1)	NC		A	1:1 pixels
54.				origin (7.4.2.2)	NC		A	top left is 0,0
55.			Color Profile	Density (7.4.3.1)	NC		A	7 bits dynamic range in gray
56.				Color Sat (7.4.3.2)	NC		A	7 bits dynamic once in grayscale
57.				Color space (7.4.3.3)	NC		24 bit RGB	Option a, reported in color space field above. See Note 8
58.			Video Interlacing (7.4.4)	NC		A	Interlaced sensors are not permitted.
59.	Full Frontal (Clause 8)	Inheritance	Inheritance (8.1)	NC		A	Inherits Frontal + Basic
60.		Scene	Scene (8.2)	NC		A	Inherits Frontal + Basic
61.		Photographic	Centered Image (8.3.2)	NC		A	Nose on vertical centerline
62.			Position of Eyes (8.3.3)	NC		A	Above horizontal centerline
63.			Width of Head (8.3.4)	NC		A	See Note 7
64.			Length of Head (8.3.5)	NC		A	See Note 7
65.		Digital	Resolution (8.4.1)	NC		CC ≥ 240	See Note 7
66.		Format	Inheritance (8.5.1)	NC		A	
67.			Image Information (8.5.2)	NC		A	

END OF TABLE

Acronym		Meaning
FAC	Face Information Record	Facial header + facial info + repetition of (image info + image data)
MF	mandatory field	[FACESTD] requires a field **shall** be present in the FAC
OF	optional field	[FACESTD] allows a field to be present in record
MV	mandatory value	[FACESTD] requires a meaningful value for a field
OV	optional value	[FACESTD] allows a meaningful value or allows 0 to be used to connote "unspecified"
NC	normative content	[FACESTD] gives normative practice for PIV. Such clauses do not define a field in the FAC.

A	as required	For PIV, value or practice is as specified in [FACESTD]
MIT	mandatory at time of instantiation	For PIV, mandatory value that **shall** be determined at the time the record is instantiated and **shall** follow the practice specified in [FACESTD]
OIT	optional at time of instantiation	For PIV, optional header value that may be determined at the time the record is instantiated

NORMATIVE NOTES:

1. The length of the entire record **shall** fit within the container size limits specified in [800-73]. These limits apply to the entire CBEFF-wrapped and signed entity, not just the [FACESTD] record. Key lengths and signing algorithms are specified in [800-78]. The size of the digital signature scales with the key length; it does not scale with the size of the biometric record.

2. More than one image may be stored in the record. It may be appropriate to store several images if appearance changes over time (beard, no beard, beard) and images are gathered at re-issuance. The most recent image **shall** appear first and serve as the default provided to applications. PIV Card capacity, however, will limit the number of images that can be stored (usually to one).

3. The recommendation is that only one image should be stored on the PIV Card.

4. PIV facial images **shall** conform to the Full Frontal Image Type defined in Clause 8 of [FACESTD].

5. Facial image data **shall** be formatted in either of the compression formats enumerated in Clause 6.2 of [FACESTD]. Both whole-image and single-region-of-interest (ROI) compression are permitted. This document ([800-76]) recommends that newly collected facial image should be compressed using ISO/IEC 15444 (i.e., JPEG 2000). This applies when images will be input to automated face recognition products for authentication, and when images are stored on PIV Cards. In this latter case, ROI compression should be used. The older ISO/IEC 10918 standard (i.e., JPEG) should be used only for legacy images.

6. Facial images **shall** be compressed using a compression ratio no higher than 15:1. However, when facial images are stored on PIV Cards JPEG 2000 should be used with ROI compression. The innermost region should be centered on the face and compressed at no more than 24:1.

7. Face recognition performance is a function of the spatial resolution of the image. [FACESTD] does not specify a minimum resolution for the Full Frontal Image Type. For PIV, faces **shall** be acquired such that a 20 centimeter target placed on, and normal to, a camera's optical axis at a range of 1.5 meters **shall** be imaged with at least 240 pixels across it. This ensures that the width of the head (i.e., dimension CC in Figure 8 of [FACESTD]) **shall** have sufficient resolution for the printed face element of the PIV Card. This specification and Clause 8.3.4 of [FACESTD] implies that the image width **shall** exceed 420 pixels. This resolution specification **shall** be attained optically without digital interpolation. The distance from the camera to the subject should be greater than or equal to 1.5 meters (for distortion reasons discussed in [FACESTD, Annex A.8]). While the size specification is a minimum, larger sizes might be used but this would require greater compression to achieve the size mandated by [800-73].

8. Facial image data **shall** be converted to the sRGB color space. As stated in Clause 7.4.3.3 of [FACESTD] this requires application of the color profile associated with the camera in use.

7.3 Performance specifications for PIV operations

[FIPS] allows automated face recognition for certain authentication purposes. Automated face recognition implementations **shall** be configured according to the specifications of Section 10. This standard does not establish qualification criteria for face recognition algorithms; Agencies should consult available test reports to select capable algorithms, for example [FACEPERF].

8.1   Scope

This Section guides implementers of biometric enrollment and authentication applications that use biometric sensors.

8.2   Available specifications and standards

The Biometric Identity Assurance Services standard [BIAS] standardizes remotely invoked biometric services, particularly it defines a framework for deploying and invoking biometrics-based identity assurance capabilities that can be readily accessed using services-based frameworks (e.g., web services). Excluded from the scope is a) single platform functionality (e.g., client-side capture) and b) integration of biometric services within an authentication protocol.

NIST Special Publication 500-288 Specification for Web Services Biometric Devices [WSBD], establishes specifications for access to, and command and control of, a target biometric sensor by enrollment or recognition clients via web services. As such, it leverages formal standardization of web services, and the wide availability of infrastructure and resources supporting such, to allow PIV implementers to maximize device-interface level interoperability i.e., the ability to replace a biometric sensor with minimal specialization. PIV implementers should consider the utility of [WSBD] and its supporting tools and documents.

PIV implementers should also note the availability of standard BioAPI Function Provider Interfaces for the sensor [BIOAPI-SENS] and for archives [BIOAPI-ARCH], and the general BioAPI standards [BIOAPI, BIOAPI-GUI, BIOAPI-FF, BIOAPI-SEC].   The simpler version 1 predecessor [BIOAPI-US] also exists. These have similar goals to those of [WSBD] but take a different approach.

9.1 Scope

All PIV biometric data **shall** be embedded in a data structure conforming to Common Biometric Exchange Formats Framework [CBEFF]. This specifies that all biometric data **shall** be digitally signed and uniformly encapsulated. This covers the following static data:

— The PIV Card fingerprints mandated by [FIPS];

— The PIV Card facial image mandated by [FIPS];

— Any other biometric data that agencies elect to place on PIV Cards (e.g.,iris);

— Any biometric records that agencies elect to retain (including purely proprietary, or derivative, elements);

— Any biometric data retained by, or for, agencies or Registration Authorities.

There are three pieces of data that are exempt:

— The [EBTS] data of Section 3.4 sent for background checks[20];

— The OCC data of Section 5.4 that is stored on the PIV Card[21];

For the relevant data above, integrity **shall** be protected by pre-pending the data with a CBEFF header and appending with a signature stored in the CBEFF signature block as depicted in the linear structure of Table 13.

Table 13 – CBEFF concatenation structure

CBEFF_HEADER	CBEFF_BIOMETRIC_RECORD	CBEFF_SIGNATURE_BLOCK
Section 9.2	Sections 3.2.2, 3.3, 6.3. 6.4 and 7.2	Section 9.3
INCITS 398 5.2.1	INCITS 398 5.2.2	INCITS 398 5.2.3

9.2 The CBEFF Header

The CBEFF Header specified in Table 14 and its notes has been established by NIST as Patron Format "PIV". This format has been established as a formal Patron Format per the provisions of [CBEFF, 6.2]. It adds definitive data types and the FASC-N field mandated by [FIPS] to a subset of the fields given in Patron Format A [CBEFF, Annex A]. It exists independently of Patron Format A. All fields of the format are mandatory.

Table 14 – Patron format PIV specification

	Patron Format PIV Field (Numbers in parentheses are [CBEFF] clauses)	Length Bytes	PIV Data Type	PIV Conformance Required Value
1.	Patron Header Version (5.2.1.4)	1	UINT	0x03
2.	SBH Security Options (5.2.1.1, 5.2.1.2)	1	Bitfield	See Note 2
3.	Biometric data block (BDB) Length	4	UINT	Length, in bytes, of the biometric data CBEFF_BIOMETRIC_RECORD
4.	Signature block (SB) Length	2	UINT	Length, in bytes, of the CBEFF_SIGNATURE_BLOCK. See Note 3
5.	BDB Format Owner (5.2.1.17)	2	UINT	Some of the CBEFF Header fields in Table 14 take modality-specific values as detailed in Table 15. Table 15, row "Biometric Format Owner", identifying the standards developer
6.	BDB Format Type (5.2.1.17)	2	UINT	Some of the CBEFF Header fields in Table 14 take modality-specific values as detailed in Table 15. Table 15, row "Biometric Format Type", identifying the standard
7.	Biometric Creation Date (5.2.1.10)	8		See Note 4 for data type
8.	Validity Period (5.2.1.11)	16		See Note 5 for data type
9.	Biometric Type (5.2.1.5)	3	UINT	Some of the CBEFF Header fields in Table 14 take modality-specific values as detailed in Table 15. Table 15, row "Biometric Type" – which modality
10.	Biometric Data Type (5.2.1.7)	1	Bitfield	Some of the CBEFF Header fields in Table 14 take modality-specific values as detailed in Table 15. Table 15, row "Biometric Data Type" – what degree of processing

---

[20] The ANSI/NIST standard [AN2011] now provides for integrity protection with its own Type 98 record. This may be useful for maintenance of PIV data. See http://biometrics.nist.gov/cs_links/standard/Type_98_Best_Practice_Guidance_v1.3.pdf.

[21] For OCC, various pieces of CBEFF information are placed in the Biometric Information Template of Table 7 as standardized in [CARD-BIO Annex C.2].

11.	Biometric Data Quality (5.2.1.9)	1	SINT	[-2,100]. A value of -2 **shall** denote that assignment was not supported by the implementation; A value of -1 **shall** indicate that an attempt to compute a quality value failed. Values from 0 to 100 **shall** indicate an increased expectation that the sample will ultimately lead to a successful match.
12.	Creator (5.2.1.12)	18	Note 6	See Note 6 for data type
13.	FASC-N	25	Note 7	See Note 7 for data type
14.	Reserved for future use	4		0x00000000

Some of the CBEFF Header fields in Table 14 take modality-specific values as detailed in Table 15.

Table 15 – CBEFF content for specific modalities

Quantity	Fingerprint Images	Fingerprint Templates	Iris Images	Facial Images	Other modalities
Section	3.2.2	3.3	6	7	-
Biometric Format Owner	0x001B i.e., M1, the INCITS Technical Committee on Biometrics	0x001B i.e., M1, the INCITS Technical Committee on Biometrics	0x0101 i.e., ISO/IEC JTC 1/SC 37 Biometrics	0x001B i.e., M1, the INCITS Technical Committee on Biometrics	For other biometric data on PIV Cards, or retained by agencies, this field **shall** be assigned in accordance with [CBEFF, 5.2.1.17].
Biometric Format Type	0x0401	0x0201	0x0009	0x0501	
Biometric Type	0x000008	0x000008	0x000010	0x000002	0x0
Biometric Data Type	b001xxxxx i.e., raw	b100xxxxx i.e., processed	b010xxxxx i.e., Intermediate	b001xxxxx i.e., raw	[CBEFF, 5.2.1.7] has 3 categories for the degree biometric data has been processed.
Quality value	Quality value **shall** be Q = 20(6 - NFIQ) where NFIQ is computed using the method of [NFIQ].		See NOTE 8	[FACESTD] requires 0 which **shall** be coded here as -2. Also 0-100 values are allowed when quality assessment is available.	
	When multiple views or samples of a biometric are contained in the record the largest (i.e., best) value should be reported. For all biometric data, whether stored on a PIV Card or otherwise, the quality value **shall** be a signed integer between -2 and 100 per the text of INCITS 358.				

NORMATIVE NOTES TO Table 14 AND Table 15.

1. Unsigned integers are denoted by UINT. Signed integers are denoted by SINT. Multi-byte integers **shall** be in Big Endian byte order.

2. The security options field has two acceptable values. The value b00001101 indicates that the biometric data block is digitally signed but not encrypted; the value b00001111 indicates the biometric data block is digitally signed and encrypted. For the mandatory [MINUSTD] elements on the PIV Card the value **shall** be b00001101.

   The fourth bit (mask 0x08) is set per prior versions of this document. The third bit (mask 0x04), which in each case is set, implements the [CBEFF, 5.2.1.2] requirement that digital signature is differentiated from message authentication code. The second bit (mask 0x02) indicates the use of encryption. The first bit (mask 0x01) indicates the use of a digital signature. See [FIPS, 800-78] for specifications on digital signatures.

3. The signature **shall** be computed over the concatenated CBEFF_HEADER and CBEFF_BIOMETRIC_RECORD in Table 13. The CBEFF_HEADER is given in Table 14. This includes the signature block length (on line 4) which may not be known before the signature is computed. This problem may be solved by conducting a two phase computation: 1) a dummy SB length value is inserted, the signature is computed (and discarded), the signature length is written into the SB length field, and 2) the signature is recomputed. For some signing algorithms an iterative procedure may be necessary.

4. This date **shall** be the date the biometric sample was acquired from the subject. For processed samples (e.g.,templates) this date should be the date of acquisition of the parent sample. Creation Date **shall** be encoded in eight bytes using a binary representation of "YYYYMMDDhhmmssZ". Each pair of characters (for example, "DD") is coded in 8 bits as an unsigned integer. Thus 17:35:30 December 15, 2005 is represented as: 00010100 00000101 00001100 00001111 00010001 00100011 00011110 01011010 where the last byte is the binary representation of the ASCII character Z which is included to indicate that the time is represented in Coordinated Universal Time (UTC). The field "hh" **shall** code a 24 hour clock value.

When multiple samples (e.g., two single finger minutiae views) are included in one record (e.g., an INCITS 378 record) and the Creation Dates are different, the Creation Date **shall** be the earliest of the multiple views.

5. The Validity Period contains two dates each of which **shall** be coded according to Normative Note 4.

   a. The validity period should start at the time when the biometric data is available for use (e.g., according to policy or issuance considerations). It **shall** be no earlier than the Creation Date. Biometric applications (e.g., authentication) should respect this date.

   b. [FIPS 201-2] limits the lifetime for biometric data. For facial images, agencies might reduce those limits to eight years[22] (Creation Date to closing date), but this might vary with technical or policy factors at agency discretion. Biometric ageing is considered to be a slow continuous process. This field therefore serves as an advisory that biometric data should be re-collected from the Cardholder at the next opportunity. This date is not intended to invalidate any function of the card (see [FIPS] for that).

6. For PIV the Creator field has length 18 bytes of which the first $K \leq 17$ bytes **shall** be printable ASCII characters, and the first of the remaining 18-K **shall** be a null terminator (zero).

7. This field **shall** contain the 25 bytes of the FASC-N component of the CHUID identifier, per [800-73, 1.8.{3,4}]. The FASC-N field may be filled with zeroes in the one exceptional case where PIV registration images are being stored before a FASC-N has been assigned. In such instances, the digital signature **shall** be regenerated once the FASC-N is known.

8. Iris quality may be set to [-2,100]. Note that formal standardization of iris image properties and quality metrics are pending in the ISO/IEC 29794-6 standard [IRISQUAL] with publications expected in late 2013 or early 2014. The value -2 indicates a failure to compute, and -1 indicates no attempt to compute quality.

9.3 The CBEFF Signature Block

The CBEFF_SIGNATURE_BLOCK contains the digital signature of the biometric data and thus facilitates the verification of the integrity of the biometric data. The process of generating a CBEFF_SIGNATURE_BLOCK is described as follows. The CBEFF_SIGNATURE_BLOCK **shall** be encoded as a CMS external digital signature as defined in [RFC5652]. The digital signature **shall** be computed over the entire CBEFF structure except the CBEFF_SIGNATURE_BLOCK itself (which means that it **shall** include the CBEFF_HEADER and the biometric records). The algorithm and key size specifications for the digital signature **shall** be implemented according to [800-78].

The CMS encoding of the CBEFF_SIGNATURE_BLOCK is as a SignedData type, and **shall** include the following information:

— The message **shall** include a version field specifying version v3

— The digestAlgorithms field **shall** be as specified in [SP 800-78]

— The encapcontentInfo **shall:**
   – Specify an eContentType of id-PIV-biometricObject
   – Omit the eContent field

— If the signature on the biometric was generated with the same key as the signature on the CHUID, the certificates field **shall** be omitted

— If the signature on the biometric was generated with a different key than the signature on the CHUID, the certificates field **shall** include only a single certificate, which can be used to verify the signature in the SignerInfo field. The certificate shall be an X.509 digital signature certificate that has been issued in accordance with Section 4.2.1 of FIPS 201-2

— The crls field **shall** be omitted

— signerInfos **shall** be present and include only a single SignerInfo

— The SignerInfo **shall:**

---

[22] The eight year duration comes from a study of the effect of face ageing on recognition accuracy [FACEPERF]. Because face-based manual verification is common, this duration is adopted as a default. Iris seems to have longer-term stability [IREX-VI]. There are no published studies of long term fingerprint stability.

- Use the issuerAndSerialNumber choice for SignerIdentifier
- Specify a digestAlgorithm in accordance with [800-78]
- Include, at a minimum, the following signed attributes:
  - A MessageDigest attribute containing the hash of the concatenated CBEFF_HEADER + Biometric Record
  - A pivFASC-N attribute containing the FASC-N of the PIV Card (to link the biometric data and PIV Card)
  - An entryUUID attribute [RFC4530] containing the 16-byte representation of the UUID value that appears in the GUID data element of the PIV Card's CHUID data element.
  - A pivSigner-DN attribute containing the subject name that appears in the PKI certificate for the entity that signed the biometric data
- Include the digital signature.

10.1   Scope

This Section establishes specifications for *configuration* of deployed biometric verification algorithms. In previous sections[23], this document includes performance specifications for *qualification* of components that influence recognition outcomes. This Section establishes minimum accuracy specifications and performance parameters for components configured and used in operational PIV biometric authentication subsystems.

NOTE   [FIPS] establishes options and requirements for all PIV functions including authentication. It allows only certain modalities to be used in PIV contexts.

10.2   Approach

FIPS 140-2 establishes minimum requirements for authentication for activation of cryptographic-modules. This Section defines analogous specifications for biometric person authentication. The specifications implement the primary security objective of using biometrics as an authentication factor.

The approach is to require recognition algorithm *operating thresholds* to be set to achieve false match rates (FMR) no higher than those advanced here. These false match rates apply to zero-effort authentication, i.e., the one-to-one comparison of sample pairs from randomly selected different persons[24]. The false match criteria implement the core biometric security objectives. These are the primary interest of a security policy.

While any false match criterion can always be met by setting a stringent[25] comparison threshold, the adoption of stringent thresholds will imply elevated false rejection rates (FNMR or FRR) because of the error-rate tradeoff[26]. High false rejection rates will inconvenience legitimate users, and it is therefore imperative that biometric systems offering sufficient performance are used – see Section 10.5.

10.3   Operating threshold specification

The threshold applied to scores from the biometric comparison algorithms **shall** be set to achieve false match rates at or below the respective values in Table 16. The threshold **shall** be calibrated in tests conformant to Annex A[27]. Agencies may require lower (more secure) FMR values; particularly some implementations can attain lower false match rates.

Table 16 – Maximum allowed false match rates by modality

Modality	Authentication	False match rate	Notes
Fingerprint minutia matching	Off-card	0.001	Applies to one comparison with one finger. See NOTE 1 and Section 10.5
Fingerprint minutia matching	On-card	0.001	
Iris image matching	Off-card	0.0001	Applies to one comparison with one eye. See NOTE 1 and Section 10.5. See NOTE 2
Face image matching	Off-card	0.001	Applies to one comparison. See NOTE 1 and Section 10.5

NOTE 1   Transactional false accept rates will be higher than these values if the transaction allows presentation of more than one sample (e.g., a second finger) or biometrics (e.g., iris and face).

---

[23] Those clauses, 4.5.3 and 4.5.4 for off-card fingerprint comparison, and 5.7.3.2 and 5.7.4 for on card comparison, qualify components requiring core minutiae-based interoperable accuracy. They do this in laboratory tests. The accuracy criteria were never intended to be adopted as operational verification criteria – particularly the FMR = 0.01 threshold was instituted to bar non-interoperable minutia detection algorithms and matchers – but is not appropriate as an Agency security policy.

[24] This represents the case where a lost card is found by someone who casually attempts a biometric authentication.

[25] For fingerprints and face, industry convention is for recognition algorithms to produce similarity scores, for which higher thresholds produce fewer false matches. For iris, the convention is to produce distance or dissimilarity scores, for which lower thresholds produce fewer false matches.

[26] As required by ISO/IEC 19795 biometric testing standards, test reports almost universally show this tradeoff using detection-error tradeoff characteristics (DETs) or almost equivalently receiver operating characteristics (ROCs).

[27] Threshold calibration information is available under NIST's MINEX and IREX programs.

NOTE 2   The iris false match rate is considerably lower because iris is readily capable of achieving this lower more secure value without appreciable increases in false non-match rates [IREX-III]

10.4   Conformance to accuracy specifications

The false match rate (FMR) requirements **shall** be assured by submitting biometric comparison algorithms to the test and calibration programs given in Table 17. Those programs **shall** provide the algorithm developer with a tabulation of false match rate vs. threshold calibration.

Table 17 – Example performance test and threshold calibration programs

Modality	Authentication	Test + Calibration Program	Status
Fingerprint minutia matching	Off-card	MINEX III	This program was formerly known as Ongoing MINEX. It includes an interoperability component (Level 1 certification) and an operational support component (Level 2 certification).
	On-card	MINEX IV	This program follows the [MINEXII] protocol to implement the Level 1 and 2 components described in MINEX III.
Iris image matching	Off-card	IREX VIII	This program follows the [IREX-I] test ensuring correct generation and matching of PIV Card (Image Type 7) standard images.
Face image matching	Off-card	Agencies may reference recent test results from any source, or take a vendor-recommended value.	Examples of such tests are [FACEPERF].

The test measurements are typically obtained by running algorithms on commodity PC hardware.

Thereafter, the algorithm provider or integrator **shall** provide documented attestation that:

— All components of the recognition software (including template generation and comparison algorithms) are functionally identical to those submitted to the recognized test and calibration program. The use of recognition algorithms on other platforms, such as wall mounted embedded processors, is allowed. The algorithm provider **shall** submit the same software to the test program wherever it is ultimately installed.

— All instances of the fielded comparison algorithms are configured with an operating decision threshold that is at least as strong as that established in FMR vs. threshold calibration.

Agencies might require inspection of source code and institute appropriate controls to ensure that the source code is indeed that installed in deployed equipment.

Additionally, Agencies could elect to conduct a biometric performance test to confirm the hypothesis that the FMR is conformant to the specification of Section 10.3.

10.4.1   Use of multiple samples with fixed thresholds

The thresholds are set to target particular false match rates between single fingers, irises or faces of different individuals. However, if agency policy is to allow two fingers or eyes to be used in an authentication attempt, then false acceptance rates will typically be twice the calibrated value. However, if a system is configured to always or conditionally require multiple instances (e.g., two fingers or two eyes), then a threshold can be adopted (using different decision or fusion logic) to target a lower false accept rate. Similarly if multiple captures (e.g., of face) are allowed, false acceptance is also increased.

10.5   Agency consideration of false rejection performance

An authentication transaction may involve several core comparisons each of which will be expected to have failure rates given by false match rates (FMR, for impostors) and false non-match rates (FNMR, for genuine comparisons). These are matching error rates defined over outcomes of sample *comparisons*. Operational authentication performance is quantified in terms of both the false reject rate (FRR) and the false accept rate (FAR) which are defined over outcomes of *transactions*[28]: In PIV, FRR is the proportion of legitimate cardholders incorrectly denied access; FAR would be the proportion of impostors incorrectly allowed access. The error rates depend on a number of factors including: the environment, the number of attempts (e.g., finger placements on the sensor), the sensor itself, the quality of the PIV Card templates' parent images, the number of fingerprints or irises invoked, and the familiarity

---

[28] A transaction might include several comparisons from repeated presentations of multiple fingers or irises.

of users with the process. The use of two fingers or irises in all authentication transactions offers substantially improved performance over single-instance authentication.

This document does not establish false rejection performance criteria – how often genuine users are unable to successfully authenticate – because it does not represent a direct security objective. Agencies are cautioned that false rejection performance is operationally vital in access control applications and is achieved by using high performance cameras and algorithms, by ensuring good quality enrollment, by correct control of the environment, by adherence to enrollment specifications, by subject and operator instruction, and by subject habituation. Agencies are therefore strongly encouraged to consider:

— Establishing a policy on how many times a subject can attempt to authenticate;

— Establishing false rejection accuracy criteria against which tests and qualification procedures can be conducted;

— Referring to false rejection performance measures reported for algorithms evaluated using the IREX test and calibration procedure;

— Referring to false rejection performance measures reported for algorithms conforming to the MINEX test and calibration procedure;

— Training staff to recognize poor quality images at time of enrollment;

— Requiring the use of multiple samples (e.g., two fingers);

— Conditions under which an alternative modality for authentication (e.g., iris instead of fingerprint) might be used;

— Conditions under which an additional modality for authentication (e.g., iris and fingerprint) might be used;

— Conducting their own supplementary tests. These might be performance tests of single products or interoperability tests, and might be used to estimate application-specific performance. The execution of tests conforming to one or more parts of the ISO/IEC 19795 standard is strongly recommended because biometric testing is a specialized discipline. Particularly a number of subtleties and difficulties exist that can potentially fatally undermine a test.

This specification does not:

— Preclude agencies from establishing more stringent false match criteria. The false match criteria can always be met by setting a high (i.e., stringent) comparison threshold. However, more stringent thresholds imply elevated false rejection errors because of the error-rate tradeoff. One mitigation is to use two fingers or two eyes.

11. Conformance to this specification

11.1 Conformance

Conformance to this specification will be achieved if an implementation and its associated data records conform to normative ("**shall**") Sections 3 through 10 but not Section 8. The following text summarizes these statements.

11.2 Conformance to PIV registration fingerprint acquisition specifications

Conformance to Section 3.2 requires the use of an [EBTS, Appendix F] certified scanner to collect a full set of fingerprint images and the application of a segmentation algorithm and the [NFIQ]-based quality assurance procedure. Images **shall** be conformant to this specification if:

— The acquisition procedures of Section 3.2 are followed. This may be tested by human observation.
— The images are conformant to [FINGSTD] as profiled by Table 4 and its normative notes.

11.3 Conformance of PIV Card fingerprint template records

Conformance to Section 3.2.2 is achieved by conformance to all the normative content of the Section. This includes production of records conformant to [MINUSTD] as profiled in Section 3.2.2. Conformance **shall** be tested by inspection of the records and performing the test assertions of the "PIV Conformance" column of Table 6. Performance certification according to Section 4.5.2.1 is necessary.

11.4 Conformance of PIV registration fingerprints retained by agencies

Conformance to Section 3.3 is achieved by conformance to all the normative content of the Section. This includes production of records conformant to [FINGSTD] as profiled in Section 3.3. Conformance **shall** be tested by inspection of the records and performing the test assertions of the "PIV Conformance" column of Table 4. Quality values [NFIQ] **shall** be checked against the NIST reference implementation.

11.5 Conformance of PIV background check records

Conformance to Section 3.4 is achieved by conformance to all the normative content of the Section. This necessitates conformance to the normative requirements of the FBI for background checks. These **shall** be tested by inspection of the transactions submitted to the FBI. This inspection may be performed either by capturing the transactions at the submitting agency or at the FBI.

11.6 Conformance to PIV authentication fingerprint acquisition specifications

Conformance to Section 4.7 **shall** be achieved if certification according to [SINGFING] is achieved, and if the resolution and area specifications are met. The [SINGFING] certification process entails inspection of output images.

11.7 Conformance of PIV facial image records

Conformance to Section 7 **shall** be achieved by conformance to all the normative content of the Section. This includes production of records conformant to [FACESTD] as profiled in Section 7.2. Conformance **shall** be tested by inspection of records and performing the test assertions of the "PIV Conformance" column of Table 12.

11.8 Conformance of CBEFF wrappers

A PIV implementation will be conformant to Section 9 if all biometric data records, whether or not mandated by this document or [FIPS], are encapsulated in conformant CBEFF records. CBEFF records **shall** be conformant if:

— The fields of the Table 14 header are present;
— The fields of Table 14 contain the allowed values as governed by its normative notes;
— A digital signature conformant to [800-78] is present;
— The values are consistent with the enclosed biometric data and the trailing digital signature.

An application that tests conformance of PIV biometric data **shall** be provided with appropriate keys to decrypt and check the digital signature.

Citation	Document
800-73	NIST Special Publication 800-73-4, Interfaces for Personal Identity Verification. There are currently three parts: Pt. 1- PIV Card Application Namespace, Data Model & Representation Pt. 2- PIV Card Application Card Command Interface Pt. 3- PIV Client Application Programming Interface http://csrc.nist.gov/publications/PubsSPs.html
800-78	NIST Special Publication 800-78-4, Cryptographic Algorithms and Key Sizes for Personal Identity Verification http://csrc.nist.gov/publications/PubsSPs.html
800-85	NIST Special Publication 800-85 A-2, PIV Card Application and Middleware Interface Test Guidelines (SP800-73-3 Compliance), July 2010 NIST Special Publication 800-85 B, PIV Data Model Test Guidelines, Jul 2006 http://csrc.nist.gov/publications/PubsSPs.html
AN2011	ANSI/NIST-ITL 1-2011 – Data Format for the Interchange of Fingerprint, Facial & Other Biometric Information, NIST Special Publication 500-290, 2011. This supersedes SP 500-245. http://www.nist.gov/itl/iad/ig/ansi_standard.cfm
APP/F	See EBTS entry below.
BAZIN	A. Bazin and T. Mansfield. An investigation of minutiae interoperability. In Proc. Fifth IEEE Workshop on Automated Identification Advanced Technologies, June 2007. AUTO-ID 2007, Alghero Italy.
BIAS	Biometric Identity Assurance Services (BIAS) SOAP Profile, Version 1.0, OASIS Standard, 24 May 2012, https://www.oasis-open.org/standards#biasv1.0 NOTE: This normatively cites the INCITS 442:2010 standard for higher level requirements and architecture, biometric operations, and data element for biometrics. INCITS 442 will be succeeded by ISO/IEC 30108 now under development. A reference implementation is available here: http://nist.gov/itl/iad/ig/upload/BIAS_20110608.zip
BIOAPI	ISO/IEC 19784-1:2006[2007] BioAPI – Biometric Application Programming Interface – Part 1: BioAPI Specification http://webstore.ansi.org/
BIOAPI-ARCH	ISO/IEC 19784-2:2007[2008] Biometric Application Programming Interface (BioAPI) – Part 2: Biometric Archive Function Provider Interface http://webstore.ansi.org/
BIOAPI-GUI	ISO/IEC 19784- 1:2006/AM1-2007 [2008], Information technology - BioAPI - Biometric Application Programming Interface - Part 1: BioAPI Specification - Amendment 1: BioGUI specification
BIOAPI-FF	ISO/IEC 19784-1, Amd. 2 19784-1:2006, Amd. 2:2009 [2009] - - Information technology - Biometric application programming interface – Part 1: BioAPI specification – Amendment 2: Framework-free BioAPI http://webstore.ansi.org/
BIOAPI-SEC	ISO/IEC 19784-1, Amd. 3 19784-1:2006, Amd. 3:2010 - Information technology -Biometric application programming interface – Part 1: BioAPI specification – Amendment 3: Support for interchange of certificates and security assertions, and other security aspects http://webstore.ansi.org/
BIOAPI-SENS	BIOAPI-SFPI: ISO/IEC 19784-4:2011 Biometric Application Programming Interface (BioAPI) – Part 4: Biometric Sensor Function Provider Interface http://webstore.ansi.org/
BIOAPI-US	ANSI INCITS 358-2002 (R2007) Information technology - BioAPI Specification (Version 1.1) and its amendment ANSI INCITS 358-2002/AM1-2007 - Amendment 1: Support for Biometric Fusion.
BIOCTS	F. Podio, D. Yaga, and C. J. McGinnis, NIST/ITL CSD Conformance Test Architectures (CTA) and Test Suites (CTS) for Biometric Data Interchange Formats http://www.nist.gov/itl/csd/biometrics/biocta_download.cfm
BSI-FACE	Markus Nuppeney, Marco Breitenstein and Matthias Niesing, EasyPASS - Evaluation of face recognition performance in an operational automated border control system. BSI and Secunet, DE. http://biometrics.nist.gov/cs_links/ibpc2010/pdfs/Nuppeney_Marcus_IBPC2010_EasyPASS_Talk_Website.pdf This presentation is accompanied by a supporting paper. http://biometrics.nist.gov/cs_links/ibpc2010/pdfs/Nuppeney2_Marcus_IBPC2010_EasyPASS_Paper_final.pdf
CARD-BIO	ISO/IEC 7816-11:2004 Identification cards -- Integrated circuit cards – Part 11: Personal verification through

Citation	Document
	biometric methods http://webstore.iec.ch/preview/info_isoiec7816-11%7Bed1.0%7Den.pdf
CARD-CMD	ISO/IEC 7816-4:2005 Identification cards – Integrated circuit cards – Part 4: Organization, security and commands for interchange http://webstore.iec.ch/preview/info_isoiec7816-4%7Bed2.0%7Den.pdf
CARD-MIN	ISO/IEC 19794-2:2011 Information technology – Biometric data interchange formats – Part 2: Finger minutiae data. This standard is not INCITS 378 nor ISO/IEC 19794-2:2005.
CBEFF	INCITS 398-2005, American National Standard for Information Technology - Common Biometric Exchange Formats Framework (CBEFF) http://webstore.ansi.org
EBTS	AFIS-DOC-01078-9.1 CJIS-RS-0010 (V9.4) – Electronic Biometric Transmission Specification, Criminal Justice Information Services, Federal Bureau of Investigation, Department of Justice, May 25, 2010. Linked from here https://www.fbibiospecs.org/docs/EBTS_v9.4_FINAL_20121212_CLEAN.pdf
	Implementers should consult https://www.fbibiospecs.org/ or request the full EBTS documentation from the FBI. Cited in this document are: 　　Appendix F　　- FBI/CJIS IMAGE QUALITY SPECIFICATIONS 　　Appendix N　　- DESCRIPTORS AND FIELD EDIT SPECIFICATIONS FOR TYPE-14 LOGICAL RECORDS 　　Appendix C　　- DESCRIPTORS AND FIELD EDIT SPECIFICATIONS FOR TYPE-2 LOGICAL RECORDS Other USG agencies have their own EBTS variants. These are not relevant to PIV.
FACEPERF	P. Grother, G.W. Quinn, and P. J. Phillips. Evaluation of 2D still-image face recognition algorithms. NIST Interagency Report 7709, National Institute of Standards and Technology, August 2010. http://face.nist.gov/mbe. See also the FRVT tests published here: http://www.nist.gov/itl/iad/ig/frvt-home.cfm
FACESTD	INCITS 385-2004, American National Standard for Information Technology - Face Recognition Format for Data Interchange http://webstore.ansi.org
FINGSTD	INCITS 381-2004, American National Standard for Information Technology - Finger Image-Based Data Interchange Format http://webstore.ansi.org
FIPS	FIPS 201-2, Personal Identity Verification, National Institute of Standards and Technology, 2013. As of the publication of this document (i.e., SP 800-76-2), FIPS 201-1 remains the formal published standard. FIPS 201-2 is expected to be released in 2013. http://csrc.nist.gov/publications/PubsFIPS.html
HSPD-12	The text of HOMELAND SECURITY PRESIDENTIAL DIRECTIVE/HSPD-12 of August 27, 2004 appears in an attachment to OMB's August 5, 2005 Memorandum M-05-24 *Implementation of HSPD 12 - Policy for a Common Identification Standard for Federal Employees and Contractors* linked from http://csrc.nist.gov/drivers/documents/Presidential-Directive-Hspd-12.html and http://www.whitehouse.gov/sites/default/files/omb/memoranda/fy2005/m05-24.pdf
ICAO	ICAO Doc 9303, Machine Readable Travel Documents: Part 1- Machine Readable Passports. Volume 2 - Specifications for Electronically Enabled Passports with Biometric Identification Capability http://www.icao.int/publications/pages/publication.aspx?docnum=9303
ICS	Methods for Testing and Specification (MTS); Implementation Conformance Statement (ICS) Proforma style guide. EG 201 058 V1.2.3 (1998-04)
IDQT	D. Potter, P. Grother, E. Tabassi, *Imaging Criteria and Test Methods for Qualification of Iris Cameras*, NIST Special Publication 500-XXX. This specification is under development (July 2013) – Latest editions of the document will be published here: http://www.nist.gov/itl/iad/ig/idqt.cfm This specification has been developed by DHS Science & Technology Directorate (S&T) and NIST.
ICNIRP-LED	ICNIRP Statement on Light-Emitting Diodes, Implications for Hazard Assessment http://www.icnirp.de/documents/led.pdf
ICNIRP-BB	ICNIRP Statement on Light-Emitting Diodes, Guidelines on Limits of Exposure to Broadband Incoherent Optical Radiation, http://www.icnirp.de/documents/broadband.pdf
IECLAMP	IEC 62471 Ed. 1.0 b:2006 Photobiological safety of lamps and lamp systems, Edition: 1.0 International Electrotechnical Commission / 26-Jul-2006 / 89 pages
	This document derives some of its content from:
	Threshold Limit Values for Chemical Substances and Physical Agents & Biological Exposure Indices, 2007, ACGIH

Citation	Document
	Worldwide www.acgih.org (American Conference of Governmental Industrial Hygienists). ANSI/IESNA RP-27.1-05 Recommended Practice for Photobiological Safety for Lamps and Lamp Systems, http://webstore.ansi.org/RecordDetail.aspx?sku=ANSI%2FIESNA+RP-27.1-05
IREX-I	P. Grother, E. Tabassi, G. W. Quinn, and W. Salamon. IREX-I: Performance of Iris Recognition Algorithms on Standard Images. Technical Report NIST Interagency Report 7629, http://iris.nist.gov/irex/, October 2009.
IREX-II	E. Tabassi, P. Grother, and W. Salamon. IREX-II: Iris Quality Calibration and Evaluation (IQCE), Performance of Iris Image Quality Assessment Algorithms, NIST Interagency Report 7820, September 30, 2011
IREX-III	P. Grother, G.W. Quinn, J. Matey, M. Ngan, W. Salamon, G. Fiumara, C. Watson, Iris Exchange III, Performance of Iris Identification Algorithms, NIST Interagency Report 7836, April 9, 2012. http://iris.nist.gov/irex
IREX-IV	IREX IV Part 1: Evaluation of Iris Identification Algorithms, George W. Quinn and Patrick Grother, NIST Interagency Report XXX. July 2013. http://iris.nist.gov/irex IREX IV Part 2: Compression profiles for iris image recognition, George W. Quinn, Patrick Grother, Mei Ngan, and Nick Rymer, NIST Interagency Report XXX, August 2013 http://iris.nist.gov/irex
IREX-V	IREX V - Best Practices for Iris Image Collection, James Matey, Elham Tabassi, George W. Quinn, Patrick Grother NIST Interagency Report XXX September 2013 http://iris.nist.gov/irex
IREX-VI	IREX VI - Temporal Stability of Iris Recognition Accuracy, NIST Interagency Report XXXX P. Grother J. R. Matey E. Tabassi G. W. Quinn and M. Chumakov http://iris.nist.gov/irex
IRISSTD	ISO/IEC 19794-6:2011 Information technology – Biometric data interchange formats – Part 6: Iris image data This document revises and replaces the 2005 iris standard. Published September 29, 2011. This standard is being amended, with publication expected 2014. The amendment does two things: It establishes conformance tests for 19794-6 iris records and, in support of that, clarifies the Image Type 7 appearance requirements. In particular it emphasizes that the sclera shall be masked. The title of the amendment is: Information Technology — Biometric data interchange formats — Part 6: Iris image format - Amendment 1: Conformance testing methodologies.
IRISQUAL	ISO/IEC 29794-6:2014 (est. date) Information technology – Biometric sample quality -- Part 6: Iris image quality This standard is currently nearing completion, at the Draft IS level.
MANSFIELD	T. Mansfield et al. Research report on minutiae interoperability tests. Technical report, Minutiae Template Interoperability Testing, 2007. http://www.mtitproject.com/DeliverableD62.pdf
MINEX04	P. Grother et al., Minutiae Interoperability Exchange Test, Evaluation Report: NISTIR 7296. The report is archived here: http://www.nist.gov/customcf/get_pdf.cfm?pub_id=150619 The ongoing assessment of minutiae algorithms is housed here: http://www.nist.gov/itl/iad/ig/ominex.cfm
MINEX II	P. Grother, W. Salamon, C. Watson, M. Indovina, and P. Flanagan, MINEX II Performance of Fingerprint Match-on-Card Algorithms Phase II / III / IV Report NIST Interagency Report 7477 (Revision I+II) http://www.nist.gov/itl/iad/ig/minexii.cfm
MINUSTD	INCITS 378-2004, American National Standard for Information Technology - Finger Minutiae Format for Data Interchange http://webstore.ansi.org
NFACS	IAFIS-DOC-07054-1.0, Criminal Justice Information Services, Federal Bureau of Investigation, Department of Justice, April 2004.
NEXUS	Canada Border Services Agency. Nexus. Expedited border clearance for pre-approved travelers into Canada and United States. 2003–2013. http://www.cbsa-asfc.gc.ca/prog/nexus/menu-eng.html.
NFIQ	E. Tabassi and C. Wilson - NISTIR 7151 - Fingerprint Image Quality, NIST Interagency Report, August 2004 http://www.nist.gov/itl/iad/ig/bio_quality.cfm
NFIQ SUMMARY	E. Tabassi and P. Grother - NISTIR 7422 Quality Summarization - Recommendations on Biometric Quality Summarization across the Application Domain
PERFSCEN	ISO/IEC 19795-2:2007 Information technology – Biometric performance testing and reporting -- Part 2: Testing methodologies for technology and scenario evaluation http://webstore.ansi.org
PERFSWAP	ISO/IEC 19795-4:2008 Information Technology -- Biometric Performance Testing and Reporting – Part 4: Interoperability Performance Testing http://webstore.ansi.org
REVFACE	A. Adler, *Sample images can be independently restored from face recognition templates*, Proc. Canadian

Citation	Document
	Conference Electronic and Computer Engineering, 1163–1166 (2003)
REVFING	Jianjiang Feng, A. K. Jain, A.K. *Fingerprint Reconstruction: From Minutiae to Phase,* IEEE Transactions on Pattern Analysis and Machine Intelligence, Volume: 33 , Issue: 2, pp. 209 – 223. Feb. 2011 R. Cappelli, A. Lumini, D Maio, *Evaluating Minutiae Template Vulnerability to Masquerade Attack* IEEE Workshop on Automatic Identification Advanced Technologies, 7-8 June 2007 pp. 174 - 179
REVIRIS	S. Venugopalan, M. Savvides, *How to Generate Spoofed Irises From an Iris Code Template,* IEEE Transactions on Information Forensics and Security, Volume: 6 , Issue: 2, pp. 385 – 395, June 2011,
RF5652	Cryptographic Message Syntax (CMS), Internet Engineering Task Force, September 2009, http://tools.ietf.org/html/rfc5652
SBMOC	D. Cooper, H. Dang, P. Lee, W. MacGregor, and K. Mehta. Secure Biometric Match-on-Card Feasibility Report. Technical report, National Institute of Standards and Technology, November 2007. Published as NIST Interagency Report 7452.
SINGFING	See "Personal Identity Verification (PIV): Image Quality Specifications For Single Finger Capture Devices". https://www.fbibiospecs.org/docs/pivspec.pdf 10 July 2006
UKIRIS	Home Office UK Border Agency. UK IRIS. Automated registered passenger entry to the UK, 2005–2013. http://www.ukba.homeoffice.gov.uk/customs-travel/EnteringtheUK/usingiris.
VOCABSTD	Part 37 of the ISO/IEC 2382 Vocabulary Standard covering Biometrics was published in December 2012. http://www.iso.org/iso/home/store/catalogue_tc/catalogue_detail.htm?csnumber=55194
WSBD	Ross J. Micheals, Kevin Mangold, Matt Aronoff, Kayee Kwong, and Karen Marshall, Specification for WS-Biometric Devices (WS-BD), NIST Special Publication 500-288, Version 1, 3/27/2012, http://www.nist.gov/itl/iad/ig/bws.cfm
WSQ31	WSQ Gray-Scale Fingerprint Image Compression Specification, IAFIS-IC-0110(V3), October 4, 2010. https://www.fbibiospecs.org/docs/WSQ_Gray-scale_Specification_Version_3_1.pdf

A.1   Scope

This Annex gives normative specifications for tests used to certify implementations that generate and/or match the mandatory minutia-based biometric elements specified by [FIPS], i.e., the two fingerprint minutiae templates placed on the PIV Card for either off-card comparison, or on-card. That is, this annex regulates the test itself, and the testing laboratory, not the products under test, and the data specifications here should not be confused with those given in Section 2.3 for fielded PIV implementations.

A.2   PIV authentication

The fingerprint templates conform to [MINUSTD] as profiled in Section 3.2.2. The use cases given in [800-73, Appendix C] detail how the templates and the PIV Card are used for interoperable authentication. Authentication may involve one or both of the fingers whose data is encoded in a PIV Card record. These will be compared with newly acquired (i.e., live) fingerprint images of either or both of the primary and secondary fingers. The inclusion of the finger position in the [MINUSTD] header allows the system to prompt the user for one or more specific fingers.

Authentication performance is quantified in terms of both the false reject rate (FRR) and the false accept rate (FAR). In PIV, FRR is the proportion of legitimate cardholders incorrectly denied access; the latter would be the proportion of impostors incorrectly allowed access. The error rates depend on a number of factors including: the environment, the number of attempts (i.e., finger placements on the sensor), the sensor itself, the quality of the PIV Card templates' parent images, the number of fingerprints invoked, and the familiarity of users with the process. The use of two fingers in all authentication transactions offers substantially improved performance over single-finger authentication. The intent of the [FIPS] specification of an interoperable biometric is to support cross-vendor and cross-agency authentication of PIV Cards. This plural aspect introduces a source of accuracy variation [MINEX04] because template generators report minutia (locations) idiosyncratically.

A.3   Test overview

This Section specifies procedures for the certification of generators and matchers of [MINUSTD] templates.

Interoperability testing requires exchange of templates between products, which **shall** therefore be tested as a group. Accordingly, the testing laboratory **shall** conduct a first round of testing to establish a primary group of interoperable template generators and matchers. Certification **shall** be determined quantitatively at the conclusion of the test. Thereafter certification requires interoperability with previously certified products.

The certification procedure **shall** be conducted offline. This allows products to be certified using very large biometric data sets, in repeatable, deterministic and therefore auditable evaluations. Offline evaluation is needed to measure performance when template data is exchanged between all pairs of interoperable products. Large populations **shall** be used to quantify the effect of sample variance on performance. A template generator is logically a converter of images to templates. A template matcher logically compares one or two templates with one or two templates to produce a similarity score. Template generators and template matchers **shall** be certified separately. This aspect is instituted because:

1. Template generation is procedurally, algorithmically and physically distinct from matching.

2. Template generation is required by [FIPS], but matching is not.

3. Fingerprint template interoperability is dependent on the quality of the PIV Card templates. The full benefits of an interoperable template will not be realized if a supplier is required to produce both a high performing generator and a high performing matcher.

4. Once a template generator is certified and deployed, its templates will be in circulation. It is necessary for all matchers to be able to process these templates. Subsequent certification rounds will be complicated if generators and matchers are certified together.

Separate certification means that a supplier may submit one or more template generators and zero or more matchers for certification. Zero or more of the submitted products **shall** ultimately be certified.

This test design conforms to the provisions of the ISO/IEC 19795-4 [PERFSWAP] standard, as profiled by this document. One Section of that standard deals with blind testing. For PIV testing the template matcher **shall** not be able to discern the source of the enrollment templates.

A.3.1 Template generator

A template generator **shall** be certified as a software library. For PIV, a template generator is a library function that **shall** convert an image into a minutiae record. The input image represents a PIV enrollment plain impression. The output template represents one of the PIV Card templates. A supplier's implementation, submitted for certification, **shall** satisfy the requirements of an application programming interface (API) specification to be published by the test organizer. The API specification will require the template generator to accept image data and produce [MINUSTD] templates conformant to Table 18. Where values or practices are not explicitly stated in Table 18, the specifications of Section 4.3 and Table 6 apply (e.g., on minutiae type). The CBEFF header and CBEFF signature **shall** not be included.

The testing laboratory **shall** input images to the generator. The template generator **shall** produce a conformant template regardless of the input. Such a template may contain zero minutiae. This provision transparently and correctly accounts for failures to enroll. In a deployed system, if quality assessment or image analysis algorithms made some determination that the input was unmatchable a failure to enroll might be declared. In an offline test such a determination **shall** result in at least a template containing zero minutiae. However, because in PIV other suppliers' matchers may be capable of handling even poor templates, it is recommended that a template generator submitted for testing should deprecate any internal quality acceptance mechanism, and attempt production of a viable template.

Table 18 – INCITS 378 specification for PIV Card template generator and matcher certification

#	Clause title and/or field name (Numbers in parentheses are [MINUSTD] clause numbers)	PIV Conformance Values Allowed	Informative Remarks
1.	Format Identifier (6.4.1)	0x464D5200	i.e., ASCII "FMR\0"
2.	Version Number (6.4.2)	0x20323000	i.e., ASCII " 20\0".
3.	Record Length (6.4.3)	$26 \le L \le 800$	26 byte header, max of 128 minutiae. See row 18.
4.	CBEFF Product Identifier Owner (6.4.4)	0	
5.	CBEFF Product Identifier Type (6.4.4)	0	
6.	Capture Equipment Compliance (6.4.5)	0	
7.	Capture Equipment ID (6.4.6)	0	
8.	Size of Scanned Image in x direction (6.4.7)	MIT	
9.	Size of Scanned Image in y direction (6.4.8)	MIT	
10.	X (horizontal) resolution (6.4.9)	197	Inherited directly from input data
11.	Y (vertical) resolution (6.4.10)	197	
12.	Number of Finger Views (6.4.11)	1	
13.	Reserved Byte (6.4.12)	0	
14.	Finger Position (6.5.1.1)	MIT	Inherited directly from input data
15.	View Number (6.5.1.2)	0	
16.	Impression Type (6.5.1.3)	0 or 2	Inherited directly from input data
17.	Finger Quality (6.5.1.4)	MIT	Inherited directly from input data
18.	Number of Minutiae (6.5.1.5)	$0 \le M \le 128$	M minutiae data records follow
19.	Minutiae Type (6.5.2.1)	01b, 10b, or 00b	See Note 1 below Table 6
20.	Minutiae Position (6.5.2.2)	MIT	See Note 7 below Table 6
21.	Minutiae Angle (6.5.2.3)	MIT	See Note 8 below Table 6
22.	Minutiae Quality (6.5.2.4)	MIT	This test specification previously required minutia quality values to be zero. This requirement no longer applies. It did not and does not apply to the PIV operational specification.
23.	Extended Data Block Length (6.6.1.1)	0	No bytes **shall** be included following this field.
END OF TABLE			

Acronym		Meaning
MIT	mandatory at time of instantiation	For PIV Certification, a mandatory value that **shall** be determined at the time the record is instantiated and **shall** follow the practice specified in [FINGSTD]

A.3.2 Template matcher

A template matcher **shall** be certified as a software library. For PIV, a matcher is a software function that compares enrollment templates with authentication templates to produce a similarity score. The similarity score **shall** be an integer or real value quantity. The enrollment templates represent the PIV Card templates. The authentication templates represent those extracted from fingerprints collected in an authentication attempt. A supplier's implementation, submitted for certification, **shall** satisfy the API specification published by the test organizer.

The API specification will support at a minimum the comparison of one authentication template (from an individual's primary or secondary fingers) with one enrollment template (from the same finger of either the same person or another individual). Both templates **shall** conform to the Table 6 profile of [MINUSTD].

The test **shall** require that all invocations of the matching function **shall** yield a similarity score regardless of the input templates. Larger scores **shall** be construed as indicating higher likelihood that the input data originated from the same person. A failure or refusal to compare the inputs **shall** in all cases result in the reporting of a score. This document recommends implementers report a low score in this case.

The input [MINUSTD] enrollment templates **shall** be prepared by the test agent using software from a supplier. The input [MINUSTD] authentication templates **shall** be the output of the template generation software provided by the supplier of the matcher under test. This means that a matcher cannot be certified as a standalone item.

A.4 Test procedure

The testing laboratory **shall** publish a test specification document. This document **shall** establish deadlines for submission of products for certification.

The supplier of a template generator **shall** submit a request for certification to the testing laboratory. The testing laboratory **shall** provide a set of image samples to these suppliers. The supplier **shall** submit templates from this data to the testing laboratory. The supplier **shall** submit the template generator to the testing laboratory. The testing laboratory **shall** execute it and check that it produces identical templates to those submitted by the supplier. The testing laboratory **shall** apply a conformance assessor to the templates. The testing laboratory **shall** report to the supplier whether identical templates were produced and whether the templates are conformant to the specifications in Table 18. This validation process may be iterative.

The supplier of a template matcher **shall** submit a request for certification to the testing laboratory. The testing laboratory **shall** provide a set of samples to these suppliers. This set **shall** support debugging and **shall** consist of images representative of those collected in PIV registration. The supplier **shall** submit similarity scores from this data to the testing laboratory. The supplier **shall** submit the template matcher to the testing laboratory. The testing laboratory **shall** execute it and check that it produces identical scores to those submitted by the supplier. The testing laboratory **shall** report to the supplier the result of the check. This validation process may be iterative.

The testing laboratory **shall** apply all template generators to the first biometric sample from each member of the test corpus. The testing laboratory **shall** invoke all template matchers to compare the resulting enrollment templates with second authentication templates from each member of the corpus. The authentication template **shall** be generated by the matcher supplier's generator (i.e., not by another supplier's generator). This **shall** be done for all pair wise combinations of template generators and template matchers. The result is a set of genuine similarity scores for each combination.

The testing laboratory **shall** invoke all template matchers to compare enrollment templates with second authentication templates from members of a disjoint population. The authentication template **shall**, in all cases, be generated by the matcher supplier's generator. This **shall** be done for all pair wise combinations of template generators and template matchers. The result is a set of impostor similarity scores for each combination. The order in which genuine and impostor similarity scores are generated **shall** be randomized (i.e., it is not implied by the order of the last two paragraphs).

The testing laboratory **shall** sum the similarity score obtained from matching of the image of a primary finger with that obtained from matching of the image of a secondary finger. This sum-rule fusion represents two-finger authentication.

A.5 Determination of an interoperable group

The testing laboratory shall compute the detection error tradeoff characteristic (DET) for all pair wise combinations of the template generators and template matchers. The testing laboratory shall generate a rectangular interoperability matrix (see [PERFSWAP]). The matrix has rows corresponding to the generators and columns corresponding to the matchers. Each element of the interoperability matrix shall be the false reject rate at a fixed false accept rate. This value corresponds to one operating point on the DET. As described in Annex A.3, the DET automatically includes the effect of failure to enroll and acquire.

An interoperable group of template generators and matchers shall be established as the largest subgroup of products submitted in an initial certification round for which all elements of the interoperability sub-matrix (i.e., FRR values) are less than or equal to 0.01 at a fixed 0.01 FAR operating point. The condition that all pair wise product combinations should be below this threshold is instituted because the PIV application is intolerant of non-interoperable pairs.

A.6 The MINEX Program

NIST's Minutia Exchange program supports interoperable standardized-minutia based biometric authentication. An overview appears in Figure 6.

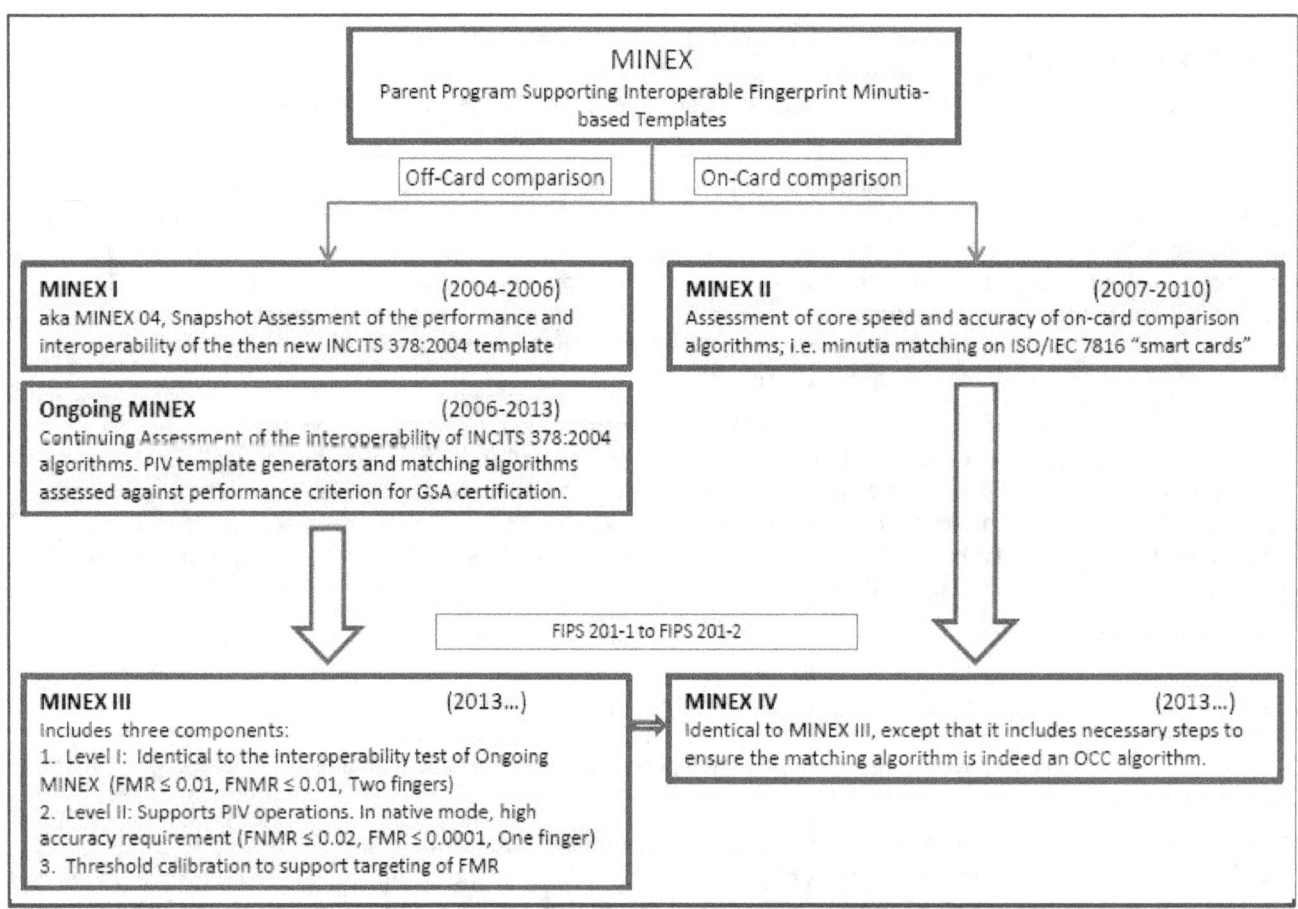

Figure 6 -- The NIST MINEX Program

This document does not establish certification criteria for iris cameras used in PIV. The motivation for this is that a formal certification procedure is not yet ready (see below) and, that as an Agency-optional biometric, iris certification costs are not warranted given the wide availability of high-performing cameras. Instead, Agencies might reasonably consider evidence of deployed track record, prior test or operational results, standards-compliance, and the guidance of the following Sub-sections.

B.1   Modes of use

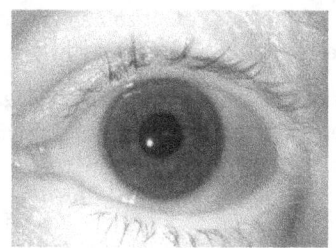

For collection of images suitable for enrollment, retention, and preparation of the specialized PIV Card images, an iris camera should routinely collect sharp, well focused, well exposed, 640x480 images in which the iris is looking at the camera directly (frontal gaze), closely centered, minimally occluded by the eyelid. One such image appears at right. Such cameras have long been available, and used for enrollment and access control.

Agencies might plan for, and select, different cameras for authentication transactions. The reason for this is that specific iris cameras are available in at least five categories: wall mounted, desktop mounted, handheld (tethered and not), stand-off, and outdoor (binocular). These variously will image just one or both eyes. Some of these specifically implement tradeoffs of speed-of-capture for image quality so that authentication can proceed more quickly than enrolment, where capture of a pristine image is needed for use over future years.

B.2   Imaging capability

Fingerprint sensors are often selected on the basis of [APP/F] or [SINGFING] certification. These are optical imaging tests designed to show that a sensor has the ability to collect fingerprint images that are faithful to the source[29]. An analogous specification for iris camera was circulated for public comment in May 2013. A joint NIST DHS S&T Special Publication titled Iris Device Qualification Test [IDQT] will result. It will define both optical imaging specifications and test methods for iris cameras. Its procedures constitute a repeatable, laboratory-based test of a camera's peak imaging capability. It is conceived of as a necessary precursor to human-in-the-loop trials that may be used to measure other performance parameters (e.g., accuracy and speed, see Section B.4).

Agencies might adopt this specification to support procurement of equipment for enrolment and authentication. It might do so by running an actual certification program, or by requiring providers to formally attest that a camera meets particular criteria.

B.3   Image quality

The ISO/IEC 29794-6 iris image quality standard is approaching completion [IRISQUAL]. That standard is likely to establish requirements on images and, separately, on iris cameras. Image quality assessment algorithms can serve as in a quality control role during applicant enrolment.

Agencies could elect to base camera selection decisions on this standard, by measuring whether a camera produces enough images of sufficiently high quality, according to the standard's metrics. This would involve capture of images from a representative human (volunteer) population. Additionally, Agencies could require that the standard's imaging specifications are met. These are a subset of the [IDQT] specifications.

B.4   Recognition performance

Agencies might elect to select cameras on the basis of a performance and usability test. Such tests are conducted to measure recognition accuracy (error rates), speed, and ease-of-use. Such tests recruit a test population to use cameras; the tests may be conducted in the laboratory or in an operational "pilot" setting. The use of a human population means the tests may be expensive to conduct, and will not be exactly repeatable.

---

[29] The [APP/F] certification originates in the need to ensure images are suitable from criminal forensic examination.

This approach measures accuracy and this necessarily involves using an iris recognition algorithm. Agencies might allow each tested camera to be associated with an algorithm, or might instead select a separate reference algorithm against which all cameras will be compared. The decision on which of these options to take will depend on the operational context and on availability of algorithms.

Agencies might choose to select cameras that demonstrate adequate accuracy and speed in a scenario test of the sort defined below. A test laboratory **should** execute such a test in formal conformance to the scenario testing requirements in Section 7 of the ISO/IEC 19795-2:2007 *Testing Methodologies for Technology and Scenario Evaluation* standard [PERFSCEN]. The test laboratory **should** additionally execute the test given the design and reporting constraints given in Table 19 - the specifications define the scenario under test, and restrict the parameters of the test design to ensure production of actionable performance data while mitigating the cost of the test.

The test laboratory **should** deliver a test report to the requesting Agency. The test report **should** conform to the reporting requirements of [PERFSCEN] and should report all accuracy and speed data mentioned in Section 6.7 of this document.

Table 19 – Profile of ISO/IEC 19795-2 for iris camera testing

#	ISO/IEC 19795-2 clause	Test parameter, topic or requirement	PIV specific scenario; test execution practice
1	7.1.2.1	Concept of operations	A test to represent a physical access control scenario for an habituated population.
2	7.1.2.2	Comparison functionality	One-to-one verification, after presentation of a PIV Card or equivalent as an identity claim. The test may proceed without reading iris imagery from the token i.e., it may be stored on a server.
3	7.1.2.3	Evaluation environment	Indoors, entrance, vestibule, atrium, or interior office, without augmentation of the environmental lighting.
4	7.1.2.4	Test platform	Not specified.
5	7.1.3.1	Test subject instruction	The test crew may be instructed on how to use the biometric system.
6	7.1.3.2	Test subject training	The test crew may execute up to ten enrollment and ten verification attempts before starting the test.
7	7.1.3.3	Attended enrollment	The enrollment attempts may be attended. The attendant should be distinct from the laboratory staff involved in the test measurements.
8	7.1.3.3	Unattended verification	The verification attempts **shall** be unattended.
9	7.1.3.4	Guidance	During enrollment, the operator may guide the user on correct preparation and use of the system.
10	7.1.3.5	Test order	The test may proceed with several devices being evaluated in parallel.
11	7.1.3.6	Test subject identifiers	The test should include presentation of a PIV Card or similar electronic token that identifies the individual.
12	7.1.4.1	Enrollment level of effort	Either or both eyes may be enrolled. The maximum number of presentations allowed for enrollment is three. The maximum duration of the entire enrollment transaction is 60 seconds.
13	7.1.4.2	Verification level of effort	Either or both eyes may be verified. The maximum number of presentations allowed for verification is three. The maximum duration of the biometric part of the entire verification transaction is 12 seconds. This may include presentation of the identity token.
14	7.1.4.3	Reference adaptation	The enrollment data **shall** not be augmented or updated during verification attempts.
15	7.1.4.5	Native configuration	The camera and ancillary software **shall** be pre-configured by the manufacturer prior to the start of the test. The test laboratory **shall** not further customize or reconfigure any component.
16	7.1.5	Multiple transactions	A test subject **shall** execute three attempts to verify as himself. This constitutes a transaction.
17	7.1.5	Multiple visits	A test subject **shall** visit on two separate days. The enrollment and genuine verification transactions **shall** not be conducted on the same day.
18	7.1.6	Executing genuine trials	A test subject **shall** execute two or more genuine transactions.
19	7.1.6	Executing impostor trials	A test subject **shall** execute at least three impostor transactions against different identities by presentation of another individual's identity token. The test subject should not be aware of whether she is making a genuine or impostor

			presentation.
20	7.1.7	Image and subject identity collection	The test laboratory **shall** retain all collected images. The camera or its ancillary software **shall** export one image per enrollment per eye in ISO/IEC 19794-6:2011 format.
21	7.2.2	Test crew habituation	The test crew should be habituated or pre-trained to mimic habituation. The test crew may have prior use of the iris camera and system.
22	7.2.3	Test crew composition	The test crew **shall** be comprised of at least 250 individuals who appear on two or more occasions. The test crew **shall** include at least 40% males. The test crew **shall** include at least 40% subjects with age above 40.
23	7.2.4	Test subject management	Each subject **shall** be assigned an identity token.
24	7.3.1	Performance	Specifications appear in Section 6.7.
25	7.3.2	Enrollment performance	Failure to enroll rate (FTE) **shall** be calculated as the fraction of persons for which at least one eye cannot be enrolled.
26	7.3.3	Failure-to-acquire performance	Failure to acquire events, if detected, **shall** be counted and reported.
27	7.3.4	Verification performance	False rejection rates **shall** be computed as the fraction of genuine subject-transactions that result in verification failure. If false acceptance occurs, testing should be stopped because false matches are unlikely to occur in a test of with this population size and the occurrence of a false match would be an indication of inappropriately weak threshold or a faulty implementation.
28	7.3.5	Identification metrics	None.
29	7.3.6	Generalized error rates including failure to acquire	Failure to acquire events encountered during genuine subject transactions **shall** be combined with false rejects to produce an effective or generalized false rejection rate.
30	7.3.7	Interim analyses	A test may be terminated early if the observed measurements support, at a statistically supported 99% confidence level, the hypothesis that the PIV requirements on FRR and capture time are violated.

Given this test report, the agency should elect to certify a camera against a set of performance requirements – an example of such follows.

EXAMPLE   The camera shall support accurate recognition. An iris camera shall be certified if it completes the performance test defined in Annex B with all of the following results:

- The proportion of subjects, executing up to three enrollment attempts, for which zero eyes can be captured i.e., failure-to-enroll rate (FTE) is at or below 0.01;

- The proportion of genuine verification transactions, each embedding up to three verification attempts, that are falsely rejected (FRR) is at or below 0.01 given a configuration consistent with false acceptance rate (FAR) at or below 0.0003 using only a PIV compliant [IRISSTD] generator and matcher;

- Retains all [IRISSTD] images to be used in offline comparisons and confirmation of the online results for which false match rate (FMR) **shall** be at or below 0.0001.

These performance specifications apply to one-to-one authentication[30].

---

[30] These performance specifications should also be suitable for one-to-many identification, which is outside of the PIV scope. However, identification requires proportionally much lower false match rates which are attainable using more stringent thresholds. These may be estimated via a calibration procedure [IREX-III].